To GEORge AND MA
ThAnks For your A
wishes.
Herb

M000291343

A Camelot of the Biomedical Sciences

The Institute Building

A Camelot of the Biomedical Sciences

RIMB Adventures

8008 Desmond Drive

Boynton Beach, Fl 33472

Copyright© 2016 by Herbert Weissbach

Printed in the United States of America

Cover & Jacket Design By Joseph Verola, http://ustarproductions.info

ISBN 978-0-9982784-0-7

First Edition

Table of Contents

Dedication

We dedicate this book to Sidney Udenfriend, John Burns and V.D. Mattia, who had the vision in 1967 to build a basic research Institute located on the Hoffmann-La Roche campus in Nutley New Jersey. This was a first for the pharmaceutical industry in the United States, and came at a time when the fields of molecular biology and molecular genetics were at such an early stage of development that one could not predict what potential there was for drug development. The concept that Udenfriend and Burns presented to Mattia, the President of Roche in the United States, could easily have been rejected, but was embraced by Mattia who convinced management in Basel that the risk was worth taking. During a period of nearly 30 years, thanks to the unwavering support of Hoffmann-La Roche, the Roche Institute of Molecular Biology, became a truly exceptional place to do science and train scientists.

Preface

What prompted us to write the story of the Roche Institute of Molecular Biology was the realization that a complete detailed history, including the training programs and the role of the Institute in shaping the future of Hoffman-La Roche, had not been done previously. An important part of the history is the unique role the Laboratory of Chemical Pharmacology in the National Heart Institute, headed by Bernard Brodie, had in the formation of the Roche Institute. It was in that NIH laboratory that four scientists, John Burns, Sidney Udenfriend, Herbert Weissbach and Alfred Pletscher initially interacted in the 1950's. A decade later their paths crossed and together they created the Roche Institute, based on their experience and the culture they were exposed to at the NIH. Of that initial group, Burns, Udenfriend and Pletscher are deceased, and we realized that if the whole story was to be told, it should be done now. With the help of David Fisher, a talented writer, who was able to take all of the information provided, and add his own research input, this book was realized. The Institute lasted for almost 30 years and for those involved from the beginning, it was the best of times, a wonderful and exciting adventure. We hope that the uniqueness of the Roche Institute still exists in the minds of all the scientists and students who trained there and are scattered in laboratories throughout the world.

Although in 1967, when the Institute was formed no one could predict the rapid advances in molecular biology that would lead to the development of the biotechnology industry, Barney Mattia, the President of Hoffmann-La Roche in Nutley, sensed that the pharmaceutical industry had to be able to move with the times. His vision is clearly seen in the following quote from a lecture he presented to Roche management in 1966 as follows:

"In the near future science and technology will advance more than in all the millennia since man's creation. We will gain the technological capacity to make starvation obsolete, to appreciably lengthen the Biblical lifespan, to change hereditary traits...There is no turning back. Science is fashioning a dramatically new international society different from any that has gone before, forged in the laboratory, marked by achievement and distinguished by change."

A year later he helped make his vision come to life.

Herbert Weissbach, Arthur Weissbach, Nathan Brot

Chapter One

In the Beginning, There Is Science

In 1953 English biologist Francis Crick wrote a brief letter to his 12-year old son, explaining as simply as possible how he and his colleague, James Watson, had changed the world forever: "Jim Watson and I have probably made a most important discovery," he wrote to Michael. "We have built a model for the structure of nucleic-acid..." DNA, he called it, a code that makes one gene different from another gene. "In other words we think we have found the basic copying mechanism by which life comes from life... You can understand that we are very excited."

For people like Crick, trying to unravel the beautiful mysteries of the universe, "the elaborate chemical mechanisms that natural selection had evolved over billions of years" was both his passion and his profession. He grew up outside the British city of Northampton dreaming of becoming a scientist. While his father ran the family's shoe factory, his grandfather was a noted amateur naturalist who corresponded with Charles Darwin and was honored by having two gastropods, snails or slugs, named after him. The young scientist began his career in physics, trying to measure the viscosity of water at extremely high temperatures, a project he later described as "the dullest problem imaginable."

That research ended abruptly during the Battle of Britain when a German bomb destroyed his laboratory.

World War II had caused governments to make huge investments in scientific research and in only a few years tremendous advances had been made: The atom had been split, penicillin had become widely available, nascent radar technology had been developed, new materials were created in laboratories, progress had been made in televising pictures; suddenly the once prosaic and even plodding world of science had become exciting. Clearly scientists were capable of creating miracle drugs – as well as really cool commercial products. Rather than pursuing product-oriented physics, Crick decided to explore the "elaborate chemical mechanisms that natural selection had evolved over billions of years" and instead became a biologist. In 1951, at the Cavendish Laboratory in Cambridge he began working with James Watson to try to uncover the fundamental secrets of life.

James Watson had grown up in Chicago. His father was a businessman; his childhood passion was birds and he intended to become an ornithologist. But in 1946, after reading Erwin Schrödinger's book What Is Life?, he became intrigued with molecular biology, especially genetics. After receiving his Ph.D. in zoology at Indiana University in the laboratory of Salvador Luria he did his postdoctoral studies in Copenhagen but then settled in

England to work with Francis Crick. By early 1953 with the help of X-ray diffraction research done by Rosalind Franklin and Maurice Wilkins, Watson and Crick had deduced the structure of DNA. They initially announced their discovery at an April, 1953 conference on proteins in Belgium; an announcement that attracted almost no attention. In fact this discovery, now widely accepted as one of the most important scientific discoveries of the 20th century wasn't even reported in the media. A month later, Watson and Crick's paper reporting their discovery was published in the journal Nature. This was the true beginning of the biotech world.

In speaking about this discovery several years later, Crick emphasized the process it took to finally reach their objective, and in doing so accurately described the thought-process familiar to every successful scientist, "The major credit I think Jim and I deserve ... is for selecting the right problem and sticking to it. It's true that by blundering about we stumbled on gold, but the fact remains that we were looking for gold. Both of us had decided, quite independently of each other, that the central problem in molecular biology was the chemical structure of the gene. ... We could not see what the answer was, but we considered it so important that we were determined to think about it long and hard, from any relevant point of view."

Watson was working in Europe because at that time there was a concentration of the scientists and technology required to attack the problem of determining the structure of DNA. The tradition of the isolated scientist pursuing knowledge simply to satisfy his own curiosity, working in a laboratory-workshop often built in a room in his own home, was already hundreds of years old on that continent. When Louis Pasteur was doing basic research into the existence of disease-causing microorganisms in the 1850's, for example, little, if any, experimental research was being done in America.

While we tend to take for granted the existence of cutting edge science laboratories in which the marvels of the world are slowly being uncovered, the concept of scientists chasing their curiosity in well-equipped labs is actually a somewhat recent phenomenon. The first chemical laboratory in America was set up at Yale in 1847, when Benjamin Silliman Jr. rented space from the college and installed his own equipment. Silliman became Yale's first Professor of Practical Chemistry in the newly created School of Applied Chemistry. As laboratory work was not considered part of a student's education, this lab was used almost exclusively by professors for their research.

It is believed that the first truly experimental medical laboratory in America was crammed into an attic at Harvard in about 1871. That appears to have been in keeping with the level of respect accorded researchers. As Harvard's leading Professor of Medicine and Botany at that time, Henry Jacob Bigelow, explained, "The excellence of the practitioner depends far more upon good judgment than great learning... and we should not, for that reason, encourage the medical student to while away his time in the labyrinths of chemistry and physiology when he ought to be learning the difference between hernia and hydrocele."

For much of American history there was little financial support available for pure research. As Nobel Prize recipient Arthur Kornberg pointed out, "The difficulty with research in our society... is the failure to understand the nature and importance of basic research. This failure can be seen among members of the lay public, political leaders, physicians, and even scientists themselves. Most people are not prepared for the time-scale of basic research and the need for a critical mass of collective effort. Fragments of knowledge (unwelcome) and unexploited are lost, as were Gregor Mendel's basic genetic discoveries. The vast majority of legislators and some scientific directors cannot accept the seeming irrelevance of basic research. Were there a record of research grants in the Stone Age, it would likely show that the major grants were awarded for proposals to build better stone axes and critics of that time ridiculed a tiny grant to someone fooling with bronze and iron. People do not realize when it comes to arguing their case for more funding, scientists who do the basic research are the least articulate, least organized, and least temperamentally equipped to justify what they are doing. In society where selling is so important, where the medium is the message, these handicaps can spell extinction."

Most of America's laboratories were built by inventors determined to transform the possibilities of science into profitable businesses. When Thomas Edison opened his Menlo Park laboratory in 1875, for example, he described it as "invention factory" from which would emerge "a minor invention every ten days and a big thing every six months or so." Almost all of the research being done at that time was consumer-oriented. There was little interest, and considerably less financial support, for medical-based research. One of the few advocates for investigation in the biological processes of life was a Baptist Minister named Frederick Gates, who wrote at the turn of the century, "I had begun to realize how woefully neglected in all civilized countries, and particularly in this one, has been the scientific study of medicine... It became clear to me that medicine could hardly hope to become a science until it should include endowed and qualified men who would give themselves to uninterrupted study and investigation, on ample salary, entirely independent of practice."

Fortunately, Reverend Gates counted among his flock one of the wealthiest men in the world, John D. Rockefeller. When Rockefeller's grandson died of scarlet fever in 1901 he was persuaded to establish one of America's first great medical research centers, the Rockefeller Institute in New York City. Within a few years its laboratories were working on the prevention and cure of contagious disease and the containment of epidemics, as well as investigating the complex biological processes critical for normal growth and development. It was Gates' vision that Rockefeller's endowment "would result in other institutions of research in a similar kind, or at least other funds for research being established, until research in this country would be conducted on a great scale, and that out of the multitudes of workers, we might be sure in the end of abundant rewards even though those rewards did not come directly from the Institute which he might found."

Without this initial support from benefactors like John D. Rockefeller, America would never have become a world leader in scientific research, because historically the federal government had showed little interest in funding investigative laboratories. The government had first allocated funds for medical research in the 1870's, when Congress created the National Board of Health to investigate the causes of deadly diseases like cholera and yellow fever. In 1887 the government opened the Hygienic Laboratory, established specifically to study bacteria, at the Marine Hospital in New York City. This one room laboratory, which officially was part of the Marine Hospital Service, is considered the beginning of what was to become one of the world's great medical research institutions, the National Institutes of Health. It wasn't much of an operation; the entire staff consisted of one full-time 'scientist,' a doctor named Joseph Kinyoun, who had been trained in the latest scientific methods.

The Marine Hospital Service had been created by Congress in 1798, when President John Adams signed a bill requiring the Treasury Department to "provide for the relief and maintenance of disabled seamen." The MHS set up a loose association of hospitals that were paid from a 20-cent a month tax levied on seamen's wages.

Beginning in the 1870's scientists in Europe had begun publishing evidence that the most fearsome infectious diseases were caused by microscopic organisms called bacteria – and people could be screened to see if they were infected. It took Dr. Kinyoun only a few months to figure out how to confirm the presence of the cholera bacillus in suspected cases. "As the (physical) symptoms... were by no means well defined," he later wrote, "the examinations were confirmatory evidence of the value of bacteria cultivation as a means of positive diagnosis."

Kinyoun demonstrated to investigators that he could positively identify the presence of these potentially devastating diseases simply by looking through his Zeiss microscope. Eventually Congress granted the Marine Health Service responsibility for examining all immigrants as they arrived on our shores for any signs of these terrible diseases; and MHS investigators set up shop on Ellis Island. Meanwhile, Dr. Kinyoun's Hygenic Laboratory was moved to Washington, D.C. in 1891. For almost a decade he remained the lab's only full-time employee. Finally, in 1901 Congress funded construction of a new building to house a laboratory where scientists could research "infectious and contagious diseases and matters pertaining to the public health."

Within a few years the Marine Health Service had been renamed the Public Health and Marine Hospital Service, which in 1912 became simply, the Public Health Service, and the Hygenic Laboratory was restructured into several different departments, which for the first time were headed by scientist-researchers rather than physicians. The laboratory also was authorized to begin researching non-contagious diseases as well as water pollution. Slowly, one small step at a time, the government was beginning to understand the importance of scientific research to protect American lives.

The successes were small, but saved lives. During World War I Hygienic Lab researchers successfully identified contaminated shaving brushes as the cause of anthrax outbreaks among troops. Other researchers discovered – before Band-Aids existed -- that the pads then being applied to protect the area around smallpox vaccinations could harbor dangerous tetanus spores. It was also during the Great War veterans of the Chemical Warfare Service attempted to secure private funding to create an Institute that would allow researchers to apply the magic of chemistry to solving medical problems. After efforts to find private sponsors had failed, Louisiana Senator Joseph Ransdell pushed through legislation in 1930 that, among other things, changed the name of the Hygienic Laboratory to the National Institute of Health (NIH), and authorized this Institute to grant fellowships to encourage research into basic biological and medical problems.

The NIH expanded slowly, moving to its current campus in Bethesda, Maryland in 1938. President Franklin Roosevelt dedicated the first buildings in 1940. Of the six original buildings on the NIH campus, most notable was Building 3. This building had been built in 1938 and originally was known as the Public Health Methods and Animal Unit Building. Like the NIH's other five original buildings, it was an attractive Georgian-style brick building that housed offices, laboratories and support space – although its attic had been transformed into an animal breeding facility. Physically there was nothing very remarkable about it; in fact researchers who worked there in the 1950s recall having to share the elevators with animals being taken to the top floor. Edward Korn remembers, "I once got into the elevator with a cow going up, who was urinating at the time. I was sort of ankle deep in it.

Among the other animals sharing Building 3 with the scientists were cats, sheep, pigs and chickens. At times seminars would be interrupted by a cackling chicken laying an egg in a nearby lab and some researchers even recalled walking out of a meeting to be greeted by eight newly born piglets.

The building presented all kinds of challenges to serious scientists. The air conditioner in the basement had to be turned off when someone was using a state-of-the-art chemical fume hood; otherwise it would carry the noxious fumes into other labs. Eventually though, Building 3 was to become legendary as a researcher's paradise, a place where scientific dreams came true. During more than six decades, before it was closed for renovation in 2001, it housed an extraordinary group of scientists. Among the people who did their work in its laboratories were three future

NIH directors, 15 future members of the National Academy of Sciences and five Nobel laureates.

In 1947 Arthur Kornberg, one of the future Nobelists, established the Section on Enzymes and Metabolism on the first floor of Building 3, which later became part of the National Institute of Arthritis and Metabolic Diseases. During Kornberg's 5 years at the NIH, before he left to take over the Chair of Microbiology at Washington University in St. Louis, the enzyme research done by Kornberg and his outstanding colleagues including Leon Heppel and Bernard Horecker gained an international reputation. By the early 1950s postdoctoral fellows were eager to join this group and talented young scientists including Bruce Ames, Gerard Hurwitz and Art Weissbach were housed in laboratories on the first floor of Building 3.

In June, 1948 President Harry Truman signed legislation creating the National Heart Institute, and officially changing the name of the NIH ever so slightly from the National Institute of Health to the National Institutes of Health. That seemingly insignificant change actually made all the difference, and the transformation of the NIH from an ordinary government laboratory into the world's center of biomedical science had begun.

Dr. James A. Shannon was named Associate Director in charge of research for the newly-formed National Heart Institute. That turned out to be an inspired choice. James Shannon was a known physiologist whose work contributed to an understanding of the fundamental processes that enabled the kidneys to concentrate and eliminate solutes like urea and creatinine.

At the start of World War II Dr. Shannon had been Director of New York University's Research Service. In 1942 the Japanese occupied Indonesia, cutting off America's main supply of quinine, at that time the world's most effective anti-malarial drug. Without a replacement drug, allied troops were vulnerable to that disease. Shannon was put in charge of a 100-bed clinical research unit at New York's Goldwater Memorial Hospital on Welfare Island and his team tested synthetic anti-malarials that remained in use years after the end of the war. The synthesis of new anti-malarial drugs was carried out at Columbia University under the supervision of Robert C. Elderfield. Among those people working with Elderfield was a young graduate student named Arthur Weissbach, who was part of the group that discovered chloroquine, one of the first synthetic anti-malarials.

After World War II James Shannon had become Director of the Squibb Institute of Medical Research, supervising the development of streptomycin, a bacterial antibiotic that proved to be the first effective drug treatment for tuberculosis. Shannon was an extraordinarily charismatic man; people used to say that all he had to do was whistle in New York and every good scientist would come running. So when the NIH's newly established National Heart Institute was searching for an Associate Director he was the perfect choice. Although the NIH offered low salaries and little prestige, within three years he had recruited a superb group of scientists. One of the people Shannon tried to recruit to the NIH was Bernard 'Steve' Brodie, who had worked with him at Goldwater Hospital during the war.

In 1950 that was not an easy decision for Brodie to make. At that time there were essentially three options available to scientists; government labs, the pharmaceutical industry or academia. Government labs were the bottom rung; almost by definition the fact that it was run by the government meant it would be a difficult place to work and you would be working with mediocre scientists. While the NIH was starting to gain some respect it still was a government lab, and was not considered a good place to build a reputation. The pharmaceutical industry was only slightly better; the pay was substantially higher but that was because they were buying your career. The path into industry was considered a one-way street, meaning it was extremely difficult to return to academia. There was little respect for corporate scientists and few prestigious labs would ever consider hiring someone who previously had "sold out" to industry. The third rung was academia, working in a university lab doing research and teaching bright young students who one day might be good enough to work in a university lab.

Steve Brodie was an NYU Professor, originally from Canada, whose pioneering work on how drugs are metabolized in the body eventually led him to be considered the founder of the field of biochemical pharmacology. Shannon's legendary persuasive powers prevailed, and Brodie agreed to set up a laboratory of Chemical Pharmacology on the second floor of Building 3. Brodie's laboratory was to become the breeding ground for the future Roche Institute of Molecular Biology. He recruited Sid Udenfriend and a technician named Julius Axelrod, who had gone to City College of New York (CCNY) and then received a Master's degree from NYU.

For Udenfriend it was a relatively easy decision; he had worked as a researcher in Brodie's NYU lab. Udenfriend had also graduated from CCNY in 1939; At that time CCNY was considered one of the great universities in America. It was supported entirely by New York City, meaning it did not charge tuition. That made all the difference to generations of smart young men and women, often the children of the immigrants who escaped persecution in

Europe, who otherwise could not have afforded college. In many ways it was considered the poor man's Harvard. As a result it was highly competitive, both to gain admission and stay there; Its Chemistry Department in particular was highly respected in the field of physiological chemistry, or as it was slowly becoming known, biochemistry.

Sid Udenfriend was one of those really smart kids who wanted to become a biochemist. After graduation he studied for his aster's degree in the Department of Biology at NYU's Graduate School. It was there he had worked with Brodie. The relationships being developed at that time, initially in university labs and soon after at the NIH, would stretch through several decades and, like ripples spreading across a great lake, make a substantial lasting impact on the entire scientific community. For example, after the war Udenfriend returned to NYU to complete his graduate studies, working for a year in the Department of Biochemistry under future Nobel Prize winner Severo Ochoa. The Spanish-born Ochoa had joined the NYU faculty in 1942 after fleeing the fascists in Spain and in 1959 received the Nobel Prize for Physiology and Medicine for his work on the synthesis of RNA.

Udenfriend received his doctorate in biochemistry from NYU in 1948, and then accepted an offer from 1947 Nobel Prize in Physiology or Medicine recipient Carl Cori to become an instructor in the Biochemistry Department at St. Louis' Washington University. Pursuing a career in science can be a very tricky road. There are no well-established paths to follow, and often personal success depends on the success of the lab in which you're working. At that point the most secure position for a biological scientist was a professorship at a notable university, while organic chemists went into private industry, usually working for one of the major pharmaceutical companies.

Shannon's most powerful recruiting tool was the future. He sold people on his vision – because he couldn't sell them on the existing facilities. In 1950 he appointed Christian Anfinsen, who would share the Nobel Prize in Chemistry in 1972, chief of the Cellular Physiology lab, located in building 3. It wasn't much of a lab; biochemist Earl Stadtman, who was brought to the NIH by Anfinsen, explains, "The laboratories in the basement of the building were just a gaping hole."

Udenfriend also was ambivalent about the move to the NIH. For him the move to Bethesda was a big gamble; In fact, Carl Cori tried to dissuade him, warning, "If you join a little known government laboratory, this will be the end of your scientific career!" But the chance to work with Brodie again proved to be irresistible – Udenfriend joined the growing parade of young scientists from Goldwater Memorial Hospital heading to

Washington. And perhaps equally important, Jim Shannon made it clear that exciting research was going to be done at NIH. He was selling his dreams. Donald Frederickson, who was to become the Director of NIH, once remembered about Shannon, "I, too, was one of those who had Jim Shannon as a spiritual father. It has often been said that Jim Shannon had an extraordinary talent for selecting people with great potential." And then he added wisely, "One hears that comment most often from people who have been selected by him."

Shannon also brought a new respect for scientific research with him to the NIH. Writing about the NIH in his book, The Social Transformation of American Medicine, author Paul Starr commends the "remarkable degree" to which "control over research was ceded to the scientific community. The approval of grant applications as well as basic policy issues rested mostly with panels of nongovernmental scientists. The individual scientist too, enjoyed autonomy with the constraints of professional competition."

While science often can be an extremely competitive world, a world in which researchers are often fighting for their share of the limited funding, Shannon made the National Heart Institute a very convivial environment. It became a place where people shared knowledge, and knew how to have a good time with their colleagues. In addition to lunch-hour stickball games and weekly interdisciplinary seminars organized by the staff, Shannon created the Applied Statistics Club, which in fact was a high stakes poker game during which the liquor Irish Mist flowed freely under the motto, "The Irish never missed!"

Among the young researchers who were recruited were brothers from the Bronx, Arthur and Herb Weissbach – although they never planned to be together at the NIH when they accepted offers to work there. Art Weissbach was five years older than his brother, and Herb spent his childhood trying to live up to his brother's reputation. They were the grandchildren of Ukrainian immigrants and the children of relatively poor but hard working first generation Americans; like so many children of that wave of immigrants their parents stressed the value of an education. While neither of their parents had gone to college they expected their sons to do so – and they not-so-secretly dreamt that their sons would become physicians. Fortunately, New York City provided tremendous free educational opportunities for those students who wanted to take advantage of them.

Both Art and Herb attended the Bronx High School of Science, even then among the best high schools in the country. Among the reasons they picked that school was that it was close enough to their apartment in the west Bronx that they could walk there without having to spend money on transportation. Neither brother went there intending to become a scientist; at that time science was generally considered to be a boring, even odd profession – and certainly not an area in which someone could earn a very good living.

During World War II everything was speeded up, so Art graduated from high school at 16 and dutifully enrolled at CCNY in pre-Med, although he had no real desire to become a physician. During his sophomore year he took a required course in organic chemistry – and in the middle of the semester he realized he had found the thing he wanted to pursue: he wanted to be a scientist. "My parents wanted me to be a doctor," he remembers, "but when I told them my intention they were mostly delighted that I was pursing an academic profession. It wasn't a profession anybody went into for the money. When I was in graduate school I thought, if I can make $10,000 a year as a scientist I'd be happy forever."

Although after graduation Art Weissbach was offered fellowships at both Berkeley and the University of Iowa, when his father became ill he was pressured to stay in New York, and to satisfy his parents enrolled in Columbia's Medical School. Columbia had an unusual structure. At that time the biochemistry department was in the medical school, because all medical students had to learn it. Ph.D. candidates were required to complete two years of medical school, at which point they could choose to pursue either an MD or Ph.D. degree or both. "After completing those two years," Art Weissbach remembers, "I dropped out of medical school and completed my Ph.D. in the Department of Biochemistry. My mother never quite understood that process. How could I drop out of medical school but still become a doctor? Years later my brother Herb once heard her trying to explain to someone what we did, 'They're like doctors,' she said, 'but not real doctors'". But what she never understood was exactly what a biochemist did. The field that was to provide the basis for molecular biology at that time was still known as biochemistry, and it was a very small field – there were fewer than 2,000 professionally registered biochemists in the nation.

Even the laboratory equipment was relatively primitive. At Columbia, for example, they had a single mass spectrometer, an essential tool for determining the chemical make-up of a substance, which members of the staff had built by hand and was run by a full-time technician. A lot of the more sophisticated glass equipment wasn't easily available, so students were taught glass blowing which enabled them to produce or replace whatever they needed. They even had to synthesize many of their own chemicals – including radioactive labeled chemicals which were very important for research.

In fact, while in graduate school Art held a part-time job preparing radioactive compounds for hospitals on the east coast. While the basic physics of radioactivity was well known, as were the potential health risks, there was considerable demand for it and pretty loose government control. Art's lab was able to purchase radioisotopes from the Atomic Energy Commission and synthesized the desired compounds. It was treated as an ordinary chemical, with vials sitting on the shelf. He explains, "As a graduate student I was doing a synthetic organic reaction to make radioactive acetic acid. We worked under a special hood in case we had an accident. And we were limited to 25 millicuries of isotope per experiment. I was making a compound for a hospital in Atlanta when something went wrong and my experiment exploded. I'd blown isotope all up and down the floor and they had to close down that part of the medical school for about a week. They waited to see if anything happened, and when nothing did they allowed me to go back in and clean it up. It was a complete mindset. But that was typical of the casual way we conducted experiments in those days."

That changed abruptly in 1949 when the Soviet Union tested its first atom bomb. Within weeks the government confiscated all the radioisotopes from university laboratories –

although several months later researchers were permitted to resume working with them.

In 1953 Art Weissbach received his PhD and began looking for a place to continue his research. Following the war many of the hugely important technological advances often made in desperation began being applied to civilian research. Advances in the use of radioisotopes, for example, had made it possible to study photosynthesis on a molecular level. So by 1952 photosynthesis was reasonably well understood, although researchers did not know what enzyme was responsible to attach carbon dioxide (CO_2) to a special sugar in the plant cell that was an essential part of the photosynthetic process. "I read an article about that problem," Art recalls, "and I thought, I know how to do that. It's simple. I learned that there was a scientist working on cellular sugar production at this still little-known government lab in Bethesda, Maryland, the NIH, named Bernard Horecker, and that I could use his technique to make this unknown sugar and find the enzyme that made the reaction possible."

The beginning of a scientific solution is the understanding of the problem. While Horecker wasn't especially intrigued by this particular problem he made a deal with Art Weissbach. Weissbach would join Horecker's lab and for six months work on whatever project he was assigned. After he proved his competency he would be permitted to work on the photosynthesis problem.

Herb Weissbach had grown up following his older brother. He attended Bronx High School of Science because of its proximity – and because he was confident he could make the school's baseball team. His dream at that point was to play for the New York Yankees and this seemed like a necessary first step. It was there he learned a little bit about science. After graduation he followed in his brother's foot steps and entered City College, once again because it was free -- and he knew he could make the baseball team. He made the freshman team but in his second year, after his father passed away, he accepted the fact that he wasn't going to play professional baseball. He needed to think about a career – and he still had his science background.

He had no great passion for science, but Bronx Science had lit a little bit of a flame for the subject so he pursued it. By then he'd spent too much of his young life being Art's younger brother and he was determined to make his own path. His brother was obtaining his graduate degree from Columbia in biochemistry so, as he remembers, "The one thing I knew for sure was that I was not going to do biochemistry. I didn't even know what it was, I just knew I wasn't going to do it. I decided to major in organic chemistry and even a future in that field was not clear to me. What changed my life was an elective course I took my senior year."

That course was Chem 59, Physiological Chemistry; it actually was another name for biochemistry, but the word 'biochemistry' was not yet widely used. The instructor was an extraordinary man named Abe Mazur, whose passion and intellect often converted dubious students into working biochemists. As another student of his, Irwin Fridovich, whose work at Duke eventually led to his being elected a Member of the National Academy of Sciences wrote, "Abe Mazur made it clear that biochemistry was the highest use of chemistry and could lead to real, testable explanations for the processes of life... and illustrated that it was possible to pursue these wonders while being paid to do so!"

Mazur had the ability to make a large body of complicated material relevant. Every human process is a chemical reaction, he taught, and to understand biology you first had to know chemistry. Then he posed the question that was to intrigue Herb Weissbach through his entire career: There are two major problems in biology, he said. "Both involve the transfer of information. The first is genetic information, how do you transfer that information from parent to child, so the new born calf looks like a calf and a human looks like a human? The second question is how does the brain process the information provided by the senses? How does the brain store and recall the information gathered by our eyes, ears, smell and touch?" These are the basic questions still asked by neuroscientists.

"He told us that in his opinion in our lifetime we will understand how the brain works but probably would not understand how genetic information was transferred" It turned out to be the complete opposite. A decade later, thanks to Abe Mazur, I was collaborating with scientists in Marshall Nirenberg's laboratory, the scientist who cracked the genetic code, breaking open the whole field of how genetic information is used by cells to make proteins. But to this day, despite the huge amount of information gained in the field of neuroscience, no one truly understands how information is stored in neurons, how it is retrieved and used to solve problems; the neuroscience code needed for information to be stored and transferred still has not been deciphered. It remains a wonderful mystery that has yet to be solved.

"I sat in that course, and for the first time in my life something clicked in my brain and I realized, this is it, this is what I want to be. Admittedly with some reluctance, it meant going into the same field as my brother. I was still following him."

Mazur was a rare teacher who not only inspired his students but then changed their lives. He had created a network of former students who were in position to hire people he recommended. His letters of recommendation were often short but to the point. What we looked for in his recommendation were the words, "He (or she) will be one of us." For example, he sent Irwin Fridovich to Duke to do postgraduate work under Philip Handler, who would later become President of the National Academy of Sciences and was awarded the National Medal of Science. Mazur put pressure on Herb Weissbach to go to graduate school, and helped get him accepted initially to the Biochemistry Department at Case Western Reserve. But Mazur had also contacted his former student Sidney Udenfriend, who was working at the NIH. Udenfriend had established a unique program that allowed graduate students to do their research at the NIH while pursuing their doctorate in biochemistry at George Washington University – and he had offered the first position in this new program to Herb Weissbach. Weissbach understood that the opportunity to be the first person in this new program to work at the NIH while pursuing his doctorate at George Washington University was exciting -- but was also a big gamble. Thus early in 1953, Herb Weissbach was faced with a significant career decision, Case Western Reserve or the NIH for graduate studies. Science is like many professional fields, the associations you make at the

34

beginning of your career can make all the difference throughout the rest of your life. This was obviously a new program and no one knew if it would work, and the NIH itself was not yet considered an especially prestigious place to do research. Even Mazur expressed some doubt about the offer to Weissbach.

But Udenfriend turned out to be a terrific salesman, convincing Weissbach to accept the offer. At that time Art Weissbach had not yet accepted his position with Horecker at the NIH. Thus, although Herb Weissbach had somewhat reluctantly followed his brother into biochemistry, he took some solace in the fact that he would end up doing his graduate work far from New York. "Finally," he remembers, "I would be away from my brother's shadow." At least he believed that until he found out later in the spring of 1953, that his brother Art would be doing postdoctoral studies also at the NIH. It turned out both of them ended up working in Building 3; and although Art was on the first floor and Herb was on the second floor, once again Herb, the graduate student, became known as Art's kid brother. Life is hard to predict. As it turned out, for a period of about 35 years, with just two exceptions, the Weissbach brothers worked in the same buildings, both at the NIH and later at the Roche Institute of Molecular Biology.

Being a student doing research at the NIH turned out to be

a once in a lifetime experience. Herb's research project was focused on a recently discovered biologically active substance called serotonin that was present in the intestine. Serotonin had been isolated, identified and crystallized in 1948 by Maurice Rapport at the Cleveland Clinic and until this point very little was known about it. Scientists still were not certain what it did, how it was made or how it metabolized. But when they put this mysterious compound on a strip of intestine, the tissue contracted, so obviously it had some effect on the digestive system. It has since been learned that serotonin has a significant role in the central nervous system, where it helps regulated mood, appetite, sex and sleep, as well as digestive processes. By 1954 it had been established by researchers that "serotonin has an important role to play in mental disorders and that a suppression of its action results in a mental disorder." It was the first step in the development of anti-depressants and antipsychotic drugs. But how it worked, what aspects of the brain it affected and how it might eventually prove valuable remained a great mystery.

Investigating how cells make and metabolize serotonin became the subject of Weissbach's doctoral thesis: "The biosynthesis and metabolism of serotonin." For him, this was a hands-on introduction to biochemistry. It turned out to be among the thousands of small steps that led eventually to the discovery of mood-altering drugs like Librium and Valium.

While doing this work Weissbach developed an assay, a method to analyze or test, for a metabolite of serotonin in urine, known as 5-hydroxyindole acetic acid, 5HIAA. This could be used to detect the presence of malignant carcinoid syndrome, an intestinal cancer, which produces large amounts of 5-HIAA. Normal patients have very low levels of 5-HIAA in their urine, but to Weissbach's surprise he had abnormally high levels of 5-HIAA in his urine compared to others in the lab. At first there was great concern that he himself might have a carcinoid tumor. However, it turned out that each morning he routinely ate a banana, which contains a high level of serotonin. This is true for several fruits and vegetables which led to the United Fruit Company giving a research award to Sid Udenfriend, and his clinical collaborator, an NIH physician Albert Sjoerdsma, for that discovery.

While working together at the NIH during this period Udenfriend and Weissbach were not only mentor and student but developed a close relationship that would last a lifetime. "It was an incredible time," Weissbach remembers. "Sid had so many ideas and I, as a student, was trying to work on all of them. I was a coauthor on close to 40 publications from my studies as a graduate student, a very productive but also a very stressful period. Like every young scientist, in the back of my mind I wondered what I was going to do next, where I was going to find a job. I was newly married and my wife Renee realized that that a career in science was not what she had expected. The combination of pressure in the lab, at school and home triggered panic attacks, and Sid wanted to help me get through this period. One day Sid sat me down and said, 'I want you to know that as long as I am able, you will always have a job.' I never forgot those words, and a decade later they would make a significant difference in my life."

As the reputation of the NIH continued to grow, a great deal of attention became focused on an extraordinary group of young scientists working in Building 3. One example is in the summer of 1953, shortly after Weissbach came to the NIH, he met Julie Axelrod, Brodie's technician. Axelrod was twice his age, and was essentially running the Brodie laboratory, although he had only a Master's degree. Weissbach told Axelrod about this new Ph.D. program that he was going to start at George Washington University Medical School in the Department of Biochemistry in the fall of 1953 and suggested Axelrod speak to George Mandel, in the Department of Pharmacology at GW. Sure enough Julie Axelrod became the second Ph.D. student to enter the Program. With a Master's degree and dozens of publications, he obtained his Ph.D. in 1955.

In 1970, only 15 years after receiving his Ph.D. Axlerod received the Nobel prize for his research in neuroscience..

Herb Weissbach survived the stress of graduate school and received his Ph.D. in 1957. During the 1950s and early 60s the Ph.D. program that Udenfriend first established had a total of 12 students, some majoring in biochemistry and others in pharmacology. In addition to Julie Axelrod two of the original students, Herb Weissbach and Darwin Prockop became members of the National Academy of Sciences. It was during this period that Herb Weissbach met another student named Ron Kuntzman, who later became an Executive Vice President of Research at Hoffmann-La Roche. Weissbach also met John Burns, who at that time was still working at Goldwater Hospital in New York City, but eventually moved his laboratory down to Brodie's lab at the NIH. John eventually became Head of Research at Hoffmann-La Roche in Nutley, N.J. in 1967.

Other talented scientists came to learn techniques in the Brodie and Udenfriend NIH laboratories. Among them was Arvid Carlsson, recipient of the Nobel Prize in 2000 for his work in Physiology or Medicine, who described Brodie in his Nobel acceptance speech as "a thrill seeker who should perhaps rightly be called 'the father of modern biochemical pharmacology,'" and explained how he made some of the discoveries that eventually led to his receiving the Nobel Award in that lab.

Also working there was Swiss neuroscientist Albert Pletscher, who had joined Hoffmann-La Roche and wanted to gain experience working with serotonin. Pletscher was a charming individual and an excellent scientist, and later became Head of Research at Roche Headquarters in Basel. Alfred Pletscher later described his time as a Visiting Scientist in Brodie's lab as "One of the highlights of my scientific career," These personal interactions that Herb Weissbach had with John Burns and Alfred Pletscher in the Brodie laboratory during that period in the 1950s were instrumental in his decision years later to help establish the Roche Institute of Molecular Biology.

The romantic image of the scientist at work is similar to that of an inventor, one man or woman who is able to single-handedly take an idea or a concept and follow it through step-by-step until it is completed. When the fundamental discoveries of science were being made in the 16th and 17th century that might have been somewhat more accurate; many scientists were curious people working alone in their own homes or with a few others, deciphering the natural forces and forging the structure of evidence-based science. But as new tools made it possible to explore microscopic worlds that began changing. By the 20th century for the most part major discoveries were actually the final step in a long, long chain that had led to that moment. Most of the experimental work being done in academia and places like the NIH was adding to the general scientific knowledge rather than changing the world. It was a step-by-small-step process in which there were a lot more 'isn't that interesting' moments than Eurekas!

As Herb Weissbach noted, "Depending on our backgrounds we approached problems in different, creative ways. Some people had a whole cell approach; they weren't interested in molecule events within the cell. They wanted to follow the morphology and changes in whole cells while others, like most of the enzymologists including myself, were interested in breaking open cells and finding out what was going on inside. It was truly the "Love of Enzymes" as Arthur Kornberg wrote. We would make an observation then begin asking questions: How does this take place, what initiates this process, what is the effect? Every question we answered raised a half-dozen additional questions that we couldn't answer."

While a scientist might dream about finding the cure for cancer, the actual work involves taking a seemingly endless series of small steps: It's a detective story, a series of small clues gathered in laboratories around the world that eventually lead to the solution. Getting the answer to a big problem required solving countless small problems. In fact, most scientists will spend much of their career trying to solve one piece of a very complicated puzzle. Sir Isaac Newton probably summed up the long road to discovery best when he said, 'If I have seen further it is by standing on the shoulders of giants."

Continued Weissbach, "The world of scientific research at that time was very different from what it is today. Most of us in that generation were interested in answering the fundamental questions of nature; we weren't thinking of practical applications. I remember being criticized by Bernie Horecker, who was on my Ph.D. committee, for describing in length in my doctoral thesis how the studies with serotonin might lead to new therapies."

That romanticized vision of science often ends with a Eureka! Moment: Thomas Watson hears Thomas Edison muffled voice demanding, "Mr. Watson, come here – I want to see you." Edward Jenner realizing previous exposure to smallpox can act as protection against the disease. Marshall Nirenberg cracking the genetic code with one experiment. It does happen, but those moments are very few and rarely reach world-changing heights. They are more like Herb Weissbach's experience: In 1958, a year after receiving his Ph.D. the NIH offered Herb the opportunity to obtain postdoctoral training and he went to Berkeley to learn microbiology from the internationally-known microbiologist Horace Barker. Barker was working on a metabolic pathway involving glutamic acid metabolism in an anaerobic organisms, a form of life that does not require oxygen to survive. This biochemical pathway had no significance in human metabolism, but was of basic interest because this metabolic pathway eventually supplied energy for this organism. Weissbach was trying to study the first reaction in this pathway, which required an enzyme and a factor that had not been identified. The factor, present in extracts of the organism was stable when initially prepared and kept in the freezer, and once the preparation was thawed it would be active in the enzymatic reaction, but quickly lost activity during the day. All attempts to prevent this loss of activity, by changing conditions continued to fail. Every day, for almost two months, Herb would start with a fresh batch in the morning, but regardless of what condition was used the factor lost activity during the day. It was incredibly frustrating.

One day Horace Barker happened to walk into the lab as Weissbach was puzzling over this problem. Barker shrugged, and in his quiet way suggested, "Maybe its light sensitive." It was one condition that Herb had not thought of. He shined a 60-watt bulb on a fresh sample for several minutes and, incredibly, it lost activity. Now it could be purified, but always in dim light; when it was obtained in larger amounts the unknown factor was orange in color, but when it was exposed to light it turned red. It became clear that the factor was a derivative of vitamin B_{12}, the first coenzyme (biologically active) form of the vitamin. Eureka! The afternoon sun that flooded the lab each day had been inactivating the factor. The lesson that Weissbach learned from that experience is that research in nature, no matter what organism is used, can lead to important discoveries.

When Herb Weissbach concluded his studies at Berkeley in 1959 he returned to Bethesda to resume work in Sid Udenfriend's laboratory – this time as an independent researcher.

The NIH in particular was becoming known as a place where scientists could do that kind of open-ended research. They could follow their own curiosity as far as it took them, without having a definitive goal or product in mind. That governing philosophy, recruit really bright people who loved the process of discovery and provide them with all the support they needed to dig deeper into the unknown world, would become the model many years later when Sid Udenfriend and John Burns, who had just become Vice-President of Research at Hoffman-La Roche in Nutley, decided in 1967 to create what was to become one of the most unique institutions associated with a pharmaceutical company and one of the top research Institutes in the world.

For people who came to the NIH from most university labs, where budgets were always a concern, it was a revelation. The government-funded labs at the NIH probably were better equipped than any other biochemistry labs in the country. In addition to the basic equipment, the beakers and the pipettes, researchers also had access to the most sophisticated instruments. But 'sophisticated instruments' of that time were still pretty rudimentary. For example, Waring blenders – direct from the household department – were used to prepare tissue extracts.

The Weissbach brothers had settled in Bethesda, Maryland at a very propitious time. Many of the labs in Building 3 moved to the newly created Clinical center (Building 10) in 1955 and 1956. More talented scientists were being recruited – and years later several of them would move to the Roche Institute. As an example Syd Spector, who later was elected President of the American Society for Pharmacology and Experimentation, remembers how Brodie would challenge the young people in his lab, "You could go to Brodie with some little bit of data and he had the faculty for taking those data and developing interesting stories. He would weave fanciful tales... (and) extrapolate your data into a global picture. And he would weave patterns for you that were incredible. Granted that many times those patterns had pores and big holes in them, I think he did that to challenge you. He would challenge us

45

by asking, 'Is that true or false?' By doing that he developed a working hypothesis to attack or to confirm and... when we left his office we would say, damn it, I don't know if it's true. I'm going to do an experiment to either refute or confirm it. And that was his strength."

Another young scientist recruited by Herb Weissbach was Nathan Brot. They first met as students in Abe Mazur's class at City College and later at Berkeley when Weissbach was doing his postdoctoral studies, and Brot was obtaining his Ph.D. degree.

Unlike Weissbach though, after graduating from CCNY personal issues had forced Brot to postpone his postgraduate education and instead he did research at Cornell Medical School for five years. After that he earned his Ph.D. at the University of California in Berkeley, working with Dr. I.L. Chaikoff, an early leading expert in cholesterol metabolism. By the time he was ready to apply for postdoctoral positions, Weissbach was firmly established at the NIH. "After I'd finished my Ph.D. at Berkeley," Brot remembers, "my wife implored me to look into an industrial job, mainly because I was older than most newly minted Ph.D. graduates and already had a family. I resisted that for philosophical reasons – I didn't want to do directed research. The concept of doing open-ended research really appealed to me, and the NIH had gained its reputation as a place where that could be done. I wrote Herb Weissbach about doing my postdoctoral work with him and he offered me a job."

Brot had to make a choice. The starting salary at the NIH was $10,250 – and he already had been offered $18,000 to work at a major pharmaceutical company. He elected to join Weissbach at the NIH .

A common thread among many of the young scientists Udenfriend hired was that they came from New York City and had been educated at city colleges. Ron Kuntzman, for example, enrolled at Brooklyn College intending to become a pharmacist. He was majoring in biology when a guidance counselor asked him to read a book about careers in science. According to this book, Kuntzman remembers, chemists earned about $35 a week while biologists made $30. When the guidance counselor once again asked Kuntzman what he wanted to be, he answered immediately, "I want to be a chemist!"

After graduating from Brooklyn College in 1954, Kuntzman was preparing to do his post-graduate work at Purdue when he skipped lunch one day to attend a lecture given by Udenfriend, who was recruiting students for his Ph.D. program. "Listening to Sid Udenfriend you'd think he ran the whole place," Kuntzman said. "But he made it sound like an exciting place to be if you really wanted to do good science." Kuntzman was intrigued by the possibility of attending George Washington University while working full-time at the NIH – and eventually accepted an offer to work with Syd Spector in Steve Brodie's lab. When Kuntzman started at the NIH he was assigned to work in Herb Weissbach's lab to learn how to assay for serotonin in tissues. "They were working on the effects of Serotonin on brain function when I got there," he explained. "I don't think any of us realized what an impact this was going to have on the whole study of depression. My work was finding a reliable method to measure the critical neurotransmitters dopamine and norepinephine in the presence of cell inhbitors."

Brodie and Udenfriend continued recruiting scientists with whom they had worked in the anti-malarial program at Goldwater during WWII, convincing John Burns to move his research laboratory to the NIH in the mid-1950s. Burns already had gained a reputation for his pioneering work on the biosynthesis and metabolism of vitamin C, research that would eventually prove significant in furthering the understanding of how drugs are metabolized in the body. In a 1954 article published in the American Journal of Medicine John Burns was the first person to use the now-common phrase 'nonsteroidal anti-inflammatory' to describe a class of pain relieving drugs.

The expansion of the NIH was also bringing a large number of talented scientists to many other laboratories on the Bethesda campus. Aaron Shatkin, then a young virologist whose discoveries about RNA would lead to his election to the National Academy of Science years later after he moved to the Roche Institute

Another young scientist named Marshall Nirenberg began his postdoctoral work at the NIH in the late 1950s. Crick and Watson's discovery of the double helix had opened up a whole new world. While it was known that DNA contained the genetic information, no one yet understood how that information was stored in the DNA. Also, they wondered how did the information contained in DNA direct the synthesis of the proteins that were expressed it in the human body and determined our characteristics? There was no evidence that DNA could directly communicate with proteins. By the late 1950's it was generally accepted that there was another type of derivative closely related to DNA that was found in the cells. It was called ribonucleic acid, or RNA, and the information in the DNA could be transcribed into the RNA. RNA was believed to be one of the missing links that contained the transcribed genetic information from DNA that could be used by the cells to make proteins..

In 1961, while still at the National Institute of Arthritis and Metabolic Diseases at the NIH, Nirenberg cracked that code using a synthetic RNA, and later showed that the sequence of every 3 chemicals in the RNA were the code word for each one of the twenty amino acids found in all proteins. Cells have a mechanism, called translation, to read the information in the RNA and produce a protein whose sequence of amino acids has been determined by the sequence of the 3 letter words in the RNA. It was like deciphering the Rosetta Stone; once he decoded the first word it was possible to determine the rest of the language. Nirenberg proved that unlike DNA, RNA consists of single strands containing 64 different code words, the language of life.

Although initially he did his research at the National Institute of Arthritis and Metabolic Diseases, his wife Perola worked as a technician in Sid Udenfriend's lab. In 1962, shortly after cracking the code, Nirenberg was poised to leave the Institute because he needed more space and personnel. There was growing competition to identify the code words of the genetic language, mainly from the laboratories of Severo Ochoa at NYU and Gobind Khorana at MIT. Nirenberg had been offered a job in academia, a job that came with a larger salary and considerably more lab space, and was seriously considering accepting it. When Perola told Udenfriend that Marshall might be leaving the NIH, he asked Herb Weissbach, and others in his lab to give up some of their lab space to keep Nirenberg there. Everyone agreed and with the support from Bob Berliner, Head of the National Heart Institute, Nirenberg moved into Udenfriend's space -- and while working there continued his studies on deciphering the code.

That initial discovery, known as the 'poly U experiment,' which identified the first code word, might well rank as the single most important experiment in biology in the 20th century. For this discovery Nirenberg was awarded the Nobel Prize. Nirenberg's discovery indirectly impacted Weissbach's work Weissbach points out, "It was my good fortune to be working on vitamin B_{12} and another vitamin called folic acid at the time. Folic acid played a role in this process by which RNA directed the first steps in protein synthesis in the cell.

"Marshall came into my lab one day in 1964 and asked me whether I'd be willing to help him elucidate that reaction by which the first amino acid was incorporated into a protein. That changed my whole life since I began working on the process called translation. These studies introduced me to the fields of molecular genetics and molecular biology and later accounted for how the Roche Institute got its name."

Nirenberg was able to attract a large number of brilliant young physicians, many of whom had joined the Public Health service after finishing medical school to satisfy their armed forces requirement. In the 1960s the NIH seemed a better place to spend a couple of years than Vietnam. One of those scientists who joined the Nirenberg laboratory was Sidney Pestka, a Polish-born physician, who years later at the Roche Institute would become acclaimed as 'the father of interferon." Pestka was typical of the young researchers who were drawn to the freedom offered at the NIH. He joined the NIH's National Heart Institute in 1962, working in Marshall Nirenberg's lab where he also met Herb Weissbach, who was working down the hall.

Later he joined the laboratory of Robert Gallo, who would play a key role in pinpointing human immunodeficiency virus (HIV) as the infectious agent responsible for AIDs. Gallo recalled working with Pestka investigating the mechanisms by which RNA created proteins, "Whenever I ordered a compound from a pharmaceutical company Sidney would ask me how I could be sure of what I had. For example, how could I know that tryptophan, a particular radioactive amino acid, was really tryptophan? I should first subject it to several tests, he instructed, including one to determine its movement under high voltage electrophoresis. Once, after almost electrocuting myself, I asked Sidney whether perhaps we might be taking things a bit too far. How do we really know water is water? How do we really know our mothers are our mothers? Sidney was not amused."

Sid Pestka was of the first scientists Herb Weissbach approached in 1967 to come to the Roche Institute, and his work on interferon became the third biotechnology drug produced commercially by recombinant technology.

The NIH at that time arguably was the most exciting in place in the world for a scientist to be working. We were all right in the middle of the discoveries that changed everything we knew about the creation of life and how we become who we are. Shannon and a group of visionary people had created an environment that allowed open-ended investigative science to be done.

By the early 1960s, renowned cardiologist Eugene Braunwald remembers, "organized research programs, which we now take for granted in medical schools, were in full bloom at the NIH. Few other places at that time were professionalizing biomedical research as a full-time occupation... We were there to conduct research, unlike what existed at most medical schools and teaching hospitals."

Dr. Michael Brown, who was awarded a Nobel Prize in Physiology and Medicine in 1985, describes what life was like at the NIH for a young physician, "The excitement was electric... Entering the pristine, logical world of pure basic science was liberating. We learned to design simple experiments that gave precise answers. (We) succeeded in drawing back the veil of mother nature and observing a simple fundamental truth. Our lives were changed forever."

The NIH proved to be a training ground for a group of scientists who eventually would use the developing technology to begin unraveling the secrets of life that had remained hidden throughout history.

From his appointment as Director in 1955 to his retirement in 1968, James Shannon led the transformation of the NIH from a growing government research center into the world's largest sponsor of biological and medical research. During those 'golden years,' as they became known, the NIH budget increased about 20% annually, to nearly $1.4 billion, allowing Shannon to initiate a massive construction program, create new research centers and increase funding for laboratories and researchers. By the time he reached the mandatory retirement age the NIH had gained worldwide recognition as one of the leading research centers in the world, and was poised to lead this nation into the biotech revolution. The men and women who trained there eventually would become the leaders of that transformative revolution. Why would scientists want to leave this great center of research?

Chapter Two

A Dream Grows in Nutley

"There are two ways of acquiring knowledge," wrote the English philosopher Roger Bacon in his book, *On Experimental Science*, in 1268, "One through reason, the other by experiment. Argument reaches a conclusion and compels us to admit it, but it neither makes us certain nor so annihilates doubt that the mind rests calm in the intuition of truth, unless it finds this certitude by way of experience... Even if a man that has never seen fire, proves by good reasoning that fire burns, and devours and destroys things, nevertheless the mind of one hearing his arguments would never be convinced, nor would he avoid fire until he puts his hand or some combustible thing into it to prove by experiment what the argument taught. But after the fact of combustion is experienced, the mind is satisfied and lies calm in the certainty of truth. Hence argument is not enough – but experience is."

As Bacon later accurately pointed out, "(E)xperimental science is a study entirely unknown to common people." At that time science was probably more feared than respected, as people mistrusted those things beyond their understanding. It has taken almost eight centuries to progress from a belief in the powers of magic and mysticism to our acceptance and reliance on empirical evidence. The system that has enabled scientists to uncover the deepest secrets of human existence is the result of a journey of small steps. It took the evolution of many disciplines for it to finally come together: These included the universal acceptance of a set of scientific standards, the growth of the chemistry world, the creation of the modern pharmaceutical industry, the introduction of financial incentives for researchers, the implementation of a legal structure that allowed researchers the right to protect and profit from their discoveries and the development of the modern laboratory.

Our founding fathers recognized the importance of scientific research to our new nation by granting to citizens in Article 1 Section 8 of the United States Constitution the power "To promote the progress of science and useful arts, by securing for limited times to authors and inventors the exclusive right to their respective writings and discoveries." But while the right to patent a discovery was enshrined in that document, at that time in America there simply wasn't a lot of discovering going on. Creating a nation out of a rugged wilderness required focusing on the problems of daily survival; from fighting wars to finding food. Very few people had the time, the tools and the imagination to explore the wonders of science. Conducting experiments was mostly an expensive hobby. And while men like Benjamin Franklin and

Thomas Jefferson had the means to follow their curiosity; most researchers spent their time using the known principles of science to invent the practical tools necessary to explore, build and trade.

In fact, there was so little organized research being done it hardly was necessary to have a name for those people doing it until 1834, when the British philosopher and historian William Whewell created the term scientist. While there were words to describe specific fields of research like chemist, mathematician and naturalist, all those people had been grouped under the broad term, philosopher – although they were considered natural or experimental philosophers. When the membership of the British Association for the Advance of Science began complaining about the lack of proper word to describe those people who investigated the wonders of the world, Whewell responded, "There was no general term by which these gentlemen could describe themselves with reference to their pursuits. Philosophers was felt to be too wide and too lofty a term... savans was rather assuming; some ingenious gentleman proposed that, by analogy with artist, they might form scientist, and added that there could be no scruple in making free with this termination when we have such words as sciolist, economist, and atheist—but this was not generally palatable."

Several years later he added, "As we cannot use physician for a cultivator of physics, I have called him a physicist. We need very much a name to describe a cultivator of science in general. I should incline to call him a Scientist. Thus we might say, that as an Artist is a Musician, Painter or Poet, a Scientist is a Mathematician, Physicist or Naturalist."

For the most part what little laboratory research being done was limited to health and medicine. Scientists were just beginning to uncover the secrets of the human body. Biology and medicine were still in their infancy and very little was known about the structures of life. Medicines were made from available herbs and plants, based on traditional formulas that had been learned from trial and error, just as they had been for centuries. There is some evidence that our oldest cave dwelling ancestors had some knowledge of herbs, and eventually Native Americans used them to treat a variety of ailments and the shaman, or medicine-men who carried on this tradition, were respected members of every tribe. The great Greek poet Homer described Egypt as a place where "the earth, the giver of grain, bears the greatest store of drugs" and "every man is a physician," and the ancient Egyptians relied on more than 700 medicinal formulations. For several hundred years in America, medical men prepared and dispensed their own curatives.

But by the middle of the 19th century a significant change had taken place – entrepreneurs realized there was a lot of money to be made selling medicines. Perry Davis' Pain Killer, a concoction of vegetable extracts, camphor, ethyl alcohol, and opiates earned its inventor a fortune by 1850. Lydia E. Pinkham put her picture of the label of her herbal-alcoholic treatment for 'a woman's ills' in 1875 and eventually became the most widely recognized woman in America. When the temperance movement led to prohibition, alcohol-based medicines became even more popular

While the opiates and alcohol in some of these 'patent medicines' actually did provide temporary relief, their financial success depended primarily on the marketing skills of the salesmen. P.T. Barnum's Seven Sutherland Sisters Hair Grower, a mixture of alcohol, vegetable oils and water, was presented by – naturally -- the seven Sutherland sisters, each of whom had flowing tresses five feet long. Traveling medicine shows presented music, skits and performances to lure crowds, with a lecture about the benefits of the products between acts. When radio became available these lectures became known as commercials.

These mostly worthless 'cure-alls' eventually became known collectively as 'snake-oil' – and real snake-oil was said to be "the strongest and best liniment known for the cure of all pain and lameness," and was very expensive. Pharmacist Ambrose

Hunsberger described the situation as a "... disorganized system of hawking medicinal remedies... There was no method of protecting the public from fraud through control or regulation of the sale of adulterated and harmful medicinal products, and the credulous citizenry of the young nation was beguiled by every description of faker and charlatan into buying their fantastic panaceas."

What had become apparent was that fortunes could be made by manufacturing and selling medicines and treatments and an industry began taking shape. While in Europe there was a long tradition of apothecaries; shops where medicines were made, prescribed and sold, and certain treatments were administered – in fact, in many places the apothecary was the local doctor – these shops were relatively rare in the early years of this country. The 1821 convention of American apothecaries in the same Philadelphia hall in which the Declaration of Independence had been signed drew up a code of ethics for pharmacists which provided for a select committee to verify the purity, safety and effectiveness of medical formulations, in essence setting up a national system of standards. The first American pharmacy textbook, which included much of the known medical knowledge, was published in 1849.

While pharmacists literally were brewing their medicines to treat the patients who came into their shops, larger companies were being formed. As early as 1668 for example, Merck had been founded in the German city of Darmstadt when Frederick Merck acquired the existing Angel Pharmacy; but in 1827 Heinrich Merck began the transition from that single store into a research-based industrial company when he began manufacturing and selling common alkaloids including morphine, strychnine, quinine, ephedrine, and nicotine. Pfizer was founded in the United States in 1849 as a chemical company, but soon was supplying medicine, painkillers and antiseptics to the Union Army. Navy Dr. Edward Squibb was appalled by the poor quality of the drugs he was given to treat sailors during the Mexican-American War and in 1858 set up his own laboratory to produce quality drugs, and soon joined Pfizer as a major supplier to the Union Army. Squibb invented the Squibb pannier, a large wooden medicine chest containing 50 different medicines to treat battlefield casualties – including ether and chloroform, a variety of herbs, and quinine and whisky. After that war Union Colonel Eli Lilly, who had studied pharmaceutical chemistry, set up his business in 1876, becoming the first drug company to focus on research and development.

With the expansion of this new industry came a need for people trained in the esoteric art of chemistry. Chemistry, in fact, is among the first 'sciences,' and can loosely be traced back to the beginnings of civilization when men found ways of separating various ores from rocks. Early philosophers speculated about the composition of the world and the elixir of life. As long ago as 380 BC the Greek philosopher Democritus speculated that all matter was composed of indestructible and invisible units called atoms – although Aristotle disputed that claim. The many theories about the composition of the world led to alchemy, the precursor to modern science that encompasses mythology, magic, metallurgy, physics, astrology, religious beliefs, spirituality and tradition. Far from being simply the search for the mythological formula that would allow them to convert metals to gold and the secrets of eternal life, it was commonly believed that alchemists laid the foundation for modern chemistry by creating a broad research structure that eventually involved into a formalized system that included theories, laboratory techniques, an experimental method and an esoteric language. Rather than being the clownish charlatans often depicted in modern literature, alchemists were accepted throughout Europe. While many alchemists were more con artists than chemists, and made a handsome profit selling counterfeit gold -- which caused Dante to consign all alchemists to the Inferno -- in fact alchemists made important contributions to society and industry, including the testing and refining of ore, the production of gunpowder, inks, dyes, paints and cosmetics, the preparation of extracts and the distillation of liquors.

In the Islamic world, by the 9th century Jābir ibn Hayyān, who was at that time known as 'the father of modern chemistry,' had introduced a systematic approach to experimental science, and had begun analyzing the chemical make-up of many substances and produced hundreds of drugs.

But by the dawn of the Enlightenment the foundations of modern science were taking shape. Sir Francis Bacon's *The Proficience and Advancement of Learning*, published in 1605, suggested the investigative format that would lead to the scientific method. In the western world Robert Boyle, a famed 17th century alchemist, is now considered the 'father of modern chemistry'; in 1663 he was one of the founders of the The Royal Society of London for Improving Natural Knowledge, the organization that is credited with turning Bacon's proposals into the accepted scientific method. In his 1661 book, The Sceptical Chymist, Boyle insisted that scientific theories must be rigorously tested before they could be regarded as true. In that same book he speculated that every visible phenomenon was actually the result of collisons of invisible particles. High on Boyle's 'wish list' of 24 inventions that would change the world were "potent drugs to alter or exalt imagination, waking, memory and other functions and appease pain, procure innocent sleep, harmless dreams, etc."

In the 18th century alchemy slowly gave way to chemistry as scientists began discovering the chemical elements that form our world; in 1835 Swedish chemist George Brandt named the dark blue pigment he found in copper ore, cobalt. A student of Brandt's named Alex Fredrik Cronstedt discovered nickel 16 years later. In 1754 Scottish chemist Joseph Black found something he named 'fixed air,' which became better known as carbon dioxide. In 1766 an English chemist named Henry Cavendish isolated 'inflammable air,' or hydrogen. In 1773, while the seeds of political dissent were taking firm root in America, Swedish chemist <u>Carl Wilhelm Scheele</u> discovered 'fire air,' or oxygen, although he did not publicly announce it; a year later British chemist Joseph Priestley independently also discovered oxygen, which he referred to as 'dephlogisticated air.'

Among those things that came together to allow the pace of discoveries to increase rapidly was the development of a somewhat standard laboratory and the necessary tools. The word laboratory is derived from the Latin, 'laboratorium'; the first part of the word, labor, means simply to 'to exert with hardship, pain, fatigue,' hard work. The second part of the word is 'oratorium', a place of prayer, making a laboratory a place where work and prayer, knowledge and faith, came together. The word was in common use as long ago as the 14th century, although at that time it actually meant simply a task that required work to be completed. By the middle of that century the word was used mostly in monasteries to refer to a workshop, similar to scriptorium, the place where scribes copied religious documents, and dormitorium. By the 16th century laboratorium referred specifically to the places where alchemists, apothecaries and metallurgists did their work, but over time it came to mean a place where scientists – especially chemists -- used specialized tools and instruments to investigate natural phenomena and processes: A laboratory.

The remains of what is believed to be among the oldest scientific laboratories were discovered by archaeologists in a dump beneath the sacristy of a church in Austria. This Renaissance laboratory was believed to have been destroyed and buried after an earthquake. Fragments of the more than 800 artifacts used in alchemy experiments were recovered included triangular crucibles, shallow bowls, porous bone-ash cups, pear-shaped earthen pots open at both ends and a rudimentary 'still' consisting of two vessels connected by a tube, as well as numerous pieces of ceramic and glass apparatuses, furnace bricks and a variety of cutting, binding and separating tools.

Apparently it was not unusual for these early labs to be located in religious institutions; in 2005, for example, a laboratory used by Leonardo Da Vinci was discovered in previously sealed rooms of a monastery next to the Basilica of Santissima Annunziata in the middle of Florence. But over time it became common for laboratories to be attached to the homes of wealthy men, and for many of these people conducting experiments became their hobby. Robert Boyle, for example, built laboratories in both his own and his sister's mansions.

As knowledge of the complexities of the natural world grew, so did the tools of the 'philosophers.' Danish astronomer Tycho Brahe constructed a three-story 'research center' on the island of Hven in the late 1500's. An observatory occupied the top level, the mathematical laboratory was on the floor level and the alchemy lab was in the basement, reflecting Brahe's belief that the microcosm and the macrocosm were inexorably connected.

Much of what we do know of early laboratories comes from 16th century paintings, engravings and woodcuts depicting the researcher in his laboratory. These usually include an array of glass test tubes and vessels on tables, shelves and window-sills, as well as books, loose papers, pens and pencils and almost always a brick fireplace with a vent hood and bellows. Sometime during the early 19th century universities expanded from the original organizing principle of teaching students what was known to encouraging them to pursue scientific and technical research. In some instances private teaching and research laboratories initially located in instructor's homes were integrated into a university.

By the mid-1800's elaborate laboratories – including separate libraries and lecture halls -- were being built throughout Europe, sparked by the exciting possibilities of the industrial revolution. The first modern chemistry lab, which revolutionized instruction, was founded by 'the father of organic chemistry' Professor Justus von Liebig in Giessen in 1824. Nine years later a physics lab was opened at Germany's Gottingen University. The first British laboratory, the College of Chemistry, now part of the University of London, was founded in 1845. The first collegiate laboratory in the United States was established in upstate New York's Rensselaer Polytechnic Institute in 1824, when founder Amos Eaton wrote to the school's patron, Stephen Van Rensselaer, "(A)bout 30 young gentlemen have proposed spending the winter with me. This number will require considerable apparatus. As usual, I again turned my mind to the almost only patron of science in our country. If you could make up your mind in favor of furnishing apparatus to the value of about three hundred dollars (let me select it economically) I could go on and pay myself with students' fees. The apparatus to remain your property, and I would engage to keep it in good repair, and to supply all articles which happen to be broken or injured··· This will be the first attempt in the world." Eaton used that grant to set up three laboratories, a classroom, a library and a museum of specimens, then described the curriculum in the 1826 catalogue, "(S)tudents of Rensselaer School learn by giving experimental and demonstrative lectures··· After visiting a bleaching factory, he returns to the laboratory and produces the chlorine gas and experiments on it until he is familiar with all the elementary principles appertaining to that curious substance···"

Within several years laboratories were being built at universities across the country, many of them based on the German model. As Frank Whitman wrote in his 1898 article in the magazine Science entitled The Beginnings of Laboratory Teaching in America, "All these scientific departments centered at first around the chemical laboratory. This is only natural, as in that science the laboratory method of teaching was first systematically developed···"

Later, he added, "The interest in science has gradually passed far beyond the mere interest in its applications··· The striking features of the last twenty years have been··· the growth of university teaching in science, as distinguished from technical."

By 1897 the importance of the lab was described in the by-then 4000 student-large Rensselaer Polytechnic Institute's annual yearbook; "From October, when the long aprons are donned for the first time, until March, the wash bottle and evaporating dish, sulphuric acid, and H_2S are faithfully used for 3 hours and "boil, filter, wash" is the motto of the day."

The creation of corporate-sponsored laboratories lagged behind. E.I. du Pont had founded his gunpowder business in 1802, and occasionally would bring saltpeter samples to a Philadelphia chemist before accepting a delivery. But as the company grew chemicals began playing a more important role, and several chemists were hired to insure quality control, investigate complaints and improve the manufacturing process. In fact, in 1880 Lamont du Pont left to open his own chemical company to make dynamite. When a rival patented smokeless gunpowder du Pont finally understood the importance of research and patent-protected inventions, so in 1902 the company established an independent chemistry lab to develop new products. Within a decade du Pont's chemical department had grown to 120 employees, and proved vital after the end of World War I by creating new products that would be manufactured in the numerous large and suddenly-idle gunpowder factories. Du Pont's lab both invented new products and improved the European products for which the company had purchased patent rights. For example they purchased a French patent for cellophane in 1924 but within a few years had developed a moisture-proof version. But wholly original products like neoprene, a synthetic rubber, and nylon were invented in du Pont's lab.

General Electric began developing "a real scientific laboratory," as it was described in corporate memos in 1900, with the specific goal of developing new products. It hired a Professor of physical chemistry from MIT, and the staff of the lab grew from 8 people in 1901 to 102 people by 1906; among the early achievements was increasing the life-span of a carbon light bulb by 25% by baking the filament. While working in GE's lab, Irving Langmuir invented the gas-filled bulb that allowed that company to dominate the American market, and in 1932 won the Nobel prize for his research.

These early corporate labs established the dynamic that became the accepted reality: Scientists had to choose between university labs that offered social prestige, research freedom and a sense of professionalism, or the larger salaries corporate labs paid to do product-oriented research. Perhaps not surprisingly, most of those industrial labs had little interest in the natural sciences. This was the new age of invention; the dawn of the age of horseless carriages and incandescent lighting, of telephones and rudimentary flying machines. What little progress being made in the natural sciences was being done in Europe. In fact, until 1894 most of the existing American pharmaceutical companies produced large batches of their existing products in factories but did little research. That changed in 1894 after European scientists at the Berlin Institute of Infectious Diseases announced they had discovered an antitoxin to treat diphtheria; the Detroit firm of Parke-Davis quickly recruited two scientists from the University of Michigan and set up the nation's first commercial biological laboratory to produce a diphtheria antitoxin. In addition to the necessary beakers, test tubes and vessels, the lab had three horses and several guinea pigs to be used in serum testing. Three years later the first commercially-produced serum to be administered in America was given to a Parke-Davis employee. The lab continued to expand and in 1902 Park-Davis opened the pharmaceutical industry's first dedicated research center, an entire building designed specifically for researchers to follow their curiosity.

At the turn of the 20th century there was little difference between the pharmaceutical and chemical industries; the new pharmaceutical companies focused more on developing over the counter consumer products like hair gels, cod liver oil and toothpaste than on prescription medicines. That began changing early in the new century when German scientist Paul Ehrlich suggested the possibility that laboratory-created chemicals could be used to treat a wide variety of targeted diseases, a process he named chemotherapy. Ehrlich, who proved his theory in 1909 when he discovered the first effective medicinal treatment for syphilis, believed that chemicals could be used to destroy or render harmless parasites, bacteria and all of the other disease causing microbes, laying the foundation for modern biological scientific research.

Prior to the first World War most of the research into disease-causing microorganisms was done by German chemists working for German chemical companies. But when Germany was blockaded during that war American chemists were forced to replicate the procedures used to manufacture several popular drugs, including the painkillers aspirin and veronal.

The transition of American pharmaceutical companies from synthesizing known agents into biological research was strengthened in 1921 when two Canadian scientists at the University of Toronto, Frederick Banting and Charles Best, proved that insulin, which was found in extracts from the pancreas of animals, would save the lives of people with diabetes. Scientists at Eli Lilly worked with them to find a way to purify the insulin in the extract and produce it in sufficient quantities to make it an effective treatment.

Old traditions died hard though. Even after World War I about half of all medicines were made by local pharmacists following drug company formulations, and were sold without prescriptions. And the big drug companies were just beginning to devote the resources necessary to discovering new treatments for common infectious diseases and often-fatal conditions. While the concept of researching the impact of chemicals to defeat disease was exciting, only a few of those new chemical substances had proven to be any more effective than the traditional treatments that already were available.

It was the second World War that changed the industry forever. While penicillin had been discovered by Alexander Fleming in 1928, it was the further research by Howard Walter Florey and Ernst Boris Chain that showed how promising this compound could be in treating infections. However, mass producing it had languished for more than a decade. When World War 2 began the United States government forged a collaboration of several companies, including Merck, Pfizer and Squibb to finally solve that problem. These companies were chosen more for the type of equipment they already had on hand, for example Merck and Pfizer both had expertise in the fermentation process, than for their research abilities. The work of Fleming, Florey and Chain saved tens of thousands of lives during the war and in 1945 they shared the Nobel prize for their work on penicillin. These studies also led to similar research into an array of potentially life changing products, from analgesics to drugs against diseases like typhoid fever. By the time the war ended the transformation of the drug companies into biological research centers was completed, and the race had begun to discover and market new biological products.

Amidst the extraordinary upheaval caused by World War II was an understanding of the vital role the biological sciences would play in the future – both in war and in peace. The availability of penicillin was considered a modern miracle – created by science: Diseases and infections that once were life-threatening were reduced to minor inconveniences. It was that recognition that led to the government's decision to enlarge and enhance the NIH as well as the pharmaceutical companies' new emphasis on research and development.

Antibiotics, medicines that help individuals and animals fight infection by destroying bacterial microorganisms, were not new. In fact, the first antibiotic can be traced back to 500 B.C. when the Chinese used moldy soybean curds to treat boils. What was new was the discovery that natural antibiotics were growing all around the world, just waiting to be discovered and isolated. The drug companies raced to find these new antibiotics, screening tens of thousands of soil and plant samples from around different terrains, from forests and ocean beds, from mountains and valleys in hopes of discovering natural treatments. Eventually this new class of drugs, among them chlortetracycline, chloramphenicol, erythromycin, and tetracycline would become widely available. Ironically, the first person to be treated successfully with streptomycin – and the third person on whom the drug was tested – was a severely wounded G.I. named Bob Dole, who later became a United States Senator, presidential candidate and co-author of the 1980 Bayh-Dole Act, which for the first time allowed drug companies to license promising drugs created in taxpayer supported university laboratories.

When World War II began 'big Pharma' as it would eventually become known, actually was composed of hundreds of small companies spread out across the country. No company controlled more than 3% of the total market. But within a decade that had changed drastically; 15 major companies accounted for 80% of all sales – and 90% of profits. As Berkeley sociologist Peter Younkin wrote, "Firms began to invest, for the first time, in research and development laboratories, they incorporated, they hired employees with backgrounds in chemistry and they began forming relationships with research universities··· Until then," he added, "firms were less vertically-integrated and there was a clear separation between universities, where research was done, manufacturers, who produced the chemicals and the pharmaceutical houses that sold them."

When George Merck became president of his family's company he stated that it would be his mission to develop products for people, not profit. But as Irwin Lerner, the former President and CEO of Roche points out, "When you get the money people involved profits are all that count; the bottom line results. And that led from the industry's traditional attitude of searching for new products for improved health to generating as much as possible to add to the bottom line." The extraordinary profits that could be generated by a single drug turned the once somewhat sedate pharmaceutical industry into a worldwide colossus, enabling companies to build sprawling corporate campuses that included state-of-the-art research facilities – in continuing search for the next miracle drug.

Among these companies that successfully survived that transition was Hoffmann-La Roche. Hoffmann-La Roche had been founded in Basel, Switzerland in 1896. Like the other small pharma companies of that time, its products were created by extracting medicines from natural sources and synthesizing them. Its first successful products included a tremendously successful orange-flavored cough syrup marketed as Sirolin, a treatment for heart disease named Digalen, the pain-killer Pantopon and another cough syrup known as Thiocol. The success of those remedies enabled Fritz Hoffmann to add his wife's maiden name, Roche, to the company products as they expanded onto four continents. The company began selling its products in America in 1905, "The Hoffmann-La Roche Chemical Works began operations in New York City with starting capital of $25,000," Roche Nutley President V.D. Mattia remembered, "...fifteen employees, three products, small quarters, high hopes and big dreams." Hoffmann-La Roche moved out of New York and established its American corporate headquarters in Nutley, New Jersey in 1928.

Key to the success of the growing company was producing and marketing vitamins in bulk, and eventually Hoffmann-La Roche would supply as much as 70% of all the synthetic vitamins to world markets. At the beginning of World War II, with Europe already embroiled in conflict, Hoffmann-La Roche essentially relocated its corporate headquarters from Switzerland to New Jersey. Included in that move were most of the company's Jewish employees, including research chemist Dr. Leo Sternbach. At the end of that war, now wary of being too dependent on vitamin sales, Roche joined the other companies intensifying research. This led to a great diversification of products, ranging from cancer chemotherapy drugs to antidepressants. In 1957 that investment in research would result in the introduction of a class of tranquilizers, drugs that sedated patients without making them tired, known as benzodiazepines. The best known benzodiazepines were librium, valium and rohypnol. Valium would become the first medicine to generate a billion-dollars a year in sales and transformed the industry.

But all of these drugs, as Arthur Weissbach points out, were "simple organic chemistry. All the things they were doing revolved about testing or synthesizing various organic molecules in search of those proverbial silver bullets. They had little interest in DNA or molecular genetics. Roche in particular was an organic synthesis company, producing and testing 50,000 chemicals a year. They had nothing to do with molecular biology, absolutely nothing. They were making a lot of money but with simple organic products."

Incredibly, the discovery of this new class of tranquilizers was mostly the result of an extraordinarily fortuitous accident. Leo Sternbach and his wife had traveled from Switzerland to Portugal by train, going directly through Nazi-occupied France with Swiss passports which did not list nationality or religion. From Portugal they sailed to America, ending up safely in Nutley, New Jersey. In the early 1950's the success of Wallace Pharmaceutical's anti-anxiety drug miltown created a sensation. Until that point barbiturates, drugs that depress the central nervous system and cause a variety of responses ranging from mild sedation to total anesthesia, were potentially too addictive to be used widely. Miltown, the drug meprobamate, was believed to be free of those dangerous effects. Roche management asked Sternbach to modify that drug to allow them to get around Wallace's patents. But Sternbach had a different idea. Many years earlier, as a post doctoral researcher in Poland, he had helped synthesize a class of unexplored compounds known as benzheptoxdiazines that originally had been conceived as dyes, but had an interesting structure that could affect the central nervous system. He spent two years researching their biological properties; none of the 40 compounds he tested appeared to have any therapeutic value. When his supervisor told him to drop that failed project and start working on discovering new antibiotics, he added methylamine to one of them then put it aside for later testing.

Eighteen months later an assistant came upon the forgotten compound while cleaning Sternbach's lab. During that time Sternbach had spent some of his own time quietly tinkering with it and had discovered that it relaxed lab mice without sedating them. Almost as an afterthought he sent it to Roche's pharmacological research laboratory for evaluation. Researchers reported "the substance has hypnotic, sedative and antistrychnine effects in mice." After several years of testing the drug was marketed as the tranquilizer librium. Three years later the more potent tranquilizer valium was approved. For more than a decade valium was the biggest-selling drug on the market, earning more than a billion dollars annually. This 'Mother' s little helper,' as the Rolling Stones sang about valium, was America's first blockbuster drug and transformed Roche from a mid-sized company known for its vitamins into a leading pharmaceutical company. Librium and valium were considered milestones in pharmaceutical history; Roche instantly became one of the most profitable companies in the industry. Its impact on the market happened to coincide with another discovery, a discovery that would rock the world.

While it is commonly accepted that what was to become known as molecular biology began with the announcement that Watson and Crick had identified the structure of DNA, in fact scientists had been pursuing that knowledge for almost a century. The phrase itself, molecular biology, was used first in 1938 by Warren Weaver, then the Director of the Division of Natural Sciences at the Rockefeller Foundation, to encourage research into what he referred to as the intersection of chemistry, biology and physics. In fact it is generally described as the place where biochemistry and genetics meet. By the time Watson and Crick began their work much had been learned about the processes of life. Biochemists had begun looking inside the structure and function of the molecules that make up all living things, while geneticists had demonstrated the vital role of genes in the life and development of individual organisms. In fact, DNA and RNA had been discovered almost a century before Watson and Crick were able to unravel the structure of DNA. It all begins with basic biology; a cell divides and when it does the newly created cell carries all the genetic information of the original cell, they are in fact exactly alike.

As early as 1865 the Augustinian monk Gregor Mendel was uncovering the secrets of inheritance, suggesting that the recipes that make each person unique are donated in the form of pairs of genes from each parent. In the early 1870's Swiss biologist Frederick Miescher, successfully isolated a small sample of what is now known to be DNA from the sperm of a salmon. It was found only in chromosomes, which suggested it had something to do with inheritance. The concept that one gene codes for one protein was shown in 1941 by George Beadle and Edward Tatum working at Stanford. In 1944, Rockefeller Institute scientists Oswald Avery and his colleagues proved that DNA carries the genetic information.

Before Avery's work most people at that time believed that information would be found in proteins, which are more complex than anything else we knew was found in the cell. It was known that proteins were composed of long chains of 20 different amino acids in varying sequences. Avery's group showed that if you took DNA from a bacterial cell that had a certain characteristic, the outside cell wall of the bacteria had ridges rather than being smooth, and transferred this DNA to a closely related bacteria that had a smooth outer wall, some of the new cells formed would be wrinkled. Most important was that the offspring of these "transformed cells" were also wrinkled. It wasn't that surprising that something could change the morphology of a cell, but to change it in such a way that its offspring would exhibit that change, that was really something. Avery wrote a letter to his brother saying that the scientific world wouldn't believe his results, but the substance responsible for this was not proteins, but rather deoxyribonucleic acid, DNA.

Avery reasoned the scientific community would question his results because DNA seemed too simple to carry such complex information. Unlike proteins, DNA consisted of only four chemicals. That didn't seem sufficient to carry all that genetic information that needs to be transferred. Those four chemicals; adenine, cytosine, guanine and thymine, were abbreviated as A, C, G, and T, and like proteins they were carried in long and complex chains. Every strand of DNA consists of those four chemicals – and only those four chemicals -- in a sequence, but it is the sequence in which they are arranged that makes all the difference. For example, the word 'l-i-f-e,' consists of four letters arranged in an easily comprehensible sequence, but those same four letters can be rearranged to form a completely different word, for example 'f-i-l-e,' which gives those letters a completely different meaning.

Numerous small but vitally important steps led Watson and Crick to their discovery. But as important as that was, it too was only a step; DNA was the cookbook for life. But no one understood how it worked. If it were a mystery it would be like discovering the body of the board game Clue's Col. Mustard lying in the parlor with a knife in his back, it is only the beginning of a long investigation. As Herb Weissbach explains, "We only discovered that genetic information is contained in the DNA in 1944. Until then we didn't know where it was in the body; we knew about sperm and eggs, but there had to be some chemical that carried this information. DNA is that chemical, it is like a genetic cookbook containing about 20,000 genes instead of recipes. Watson and Crick worked out the unique double helical structure of DNA, which provided the basis in principle of how genetic information could be copied (replicated) when a cell divides so that the newly formed cells have the same genetic information as the parent cell. But there also had to be a way by which the information in the genes was used by the cells to make protein, which determines who you are and all of your characteristics.

The discoveries by Avery and coworkers and Watson and Crick opened up entirely new paths of investigation. It was like someone had opened a door to a new world–that would have a profound impact on the future of mankind.

A most significant question still remained. How is the information in the DNA used by each cell to determine the characteristics of the individual? The characteristics one has are due to the proteins that are made based on the information you inherited in the DNA. In some way the information in the DNA told the cells to make proteins that determined who and what you are. DNA is like the Commanding General of the Army, while the proteins are the grunts who do all the work. Proteins determine what a person looks like, how that person functions and, sometimes, when a person gets a disease. Another question also had to be answered. If all the cells of the body have the same genetic material as the fertilized egg, how does each cell know what proteins to make – give me a pair of blue eyes --to produce the right proteins to make those eyes blue? How does a liver cell produce only proteins needed for the liver to function, or a brain cell makes only brain proteins? "We knew the four letters of the language," Herb Weissbach says, "but we couldn't understand what the words of this language were or how the words were put together to convey information. It was like looking at the Rosetta Stone; how was the information contained in the sequence of the 4 chemical letters in the DNA? From the beginning of life on this planet cells knew how to read the information in their DNA and make proteins. Yet humans with their vast intellect still could not; and as the 1950s came to an end we still did not understand the language of genetics. What was the code?"

DNA, it turned out, is found mainly in the nucleus of the cell. But proteins are made outside the nucleus, in what is called the cytoplasm. Somehow that recipe, that information, has to be transmitted from the nucleus into the cytoplasm. The name of the messenger turned out to be ribonucleic acid, or RNA. The information in the DNA is rewritten, or transcribed, into the messenger RNA, and the RNA then moves into the cytoplasm and the information that it contains from the DNA can now be read by the cell. The messenger RNA has the same sequence of 4 letters as in the DNA, except that the chemical abbreviated as T has been modified and is replaced with the chemical uridine, abbreviated with the letter U.

Proteins are made from long strands of amino acids. It was known that there were 20 different amino acids and somehow the sequence of the amino acids in a protein is determined by the sequence of the 4 chemicals, originally in the DNA, that was rewritten in the RNA. In some way the amino acids have to communicate with the RNA so that the sequence of chemicals in the RNA can determine the sequence of amino acids that will be present in a specific protein. One cannot imagine a more important language to decode than the genetic language.

There are many people who believe the allies were able to win the Second World War because we successfully broke the Enigma Code, the code in which the Germans gave orders to its armies around the world. For a long time Enigma was considered an unbreakable code because it seemed to have an infinite number of possible settings. The genetic code should be simpler since it is a language with only 4 letters, but yet the best scientists in the world, with all of this accumulated brain power, could not decipher it until 1961, 17 years from the time that DNA was shown to contain the genetic information. A young scientist at the NIH, by the name of Marshall Nirenberg, cracked the code.

Marshal Nirenberg came to the NIH in 1957 as a postdoctoral fellow in the NIAMD. After a short period he obtained a permanent position in that Institute. Herb Weissbach had interactions with Marshall at one time since Marshall was interested initially in the enzyme aldehyde dehydrogenase that Herb had experience with from his earlier studies on serotonin and he helped Marshall set up the assay for that enzyme. Marshall, like so many other scientists at that time, became interested in how genetic information in the messenger RNA is used by cells to make proteins. The problem for all of the scientists interested in this field was what system could be used to identify how the information was coded in the sequence of the 4 chemical letters originally in the DNA, which was transcribed into messenger RNA. Art Weissbach, studying DNA synthesis, had set up a bacterial system, similar to what Marshall would eventually use to crack the genetic code. Herb Weissbach, working with Ed Katz, a microbiologist from Georgetown University, thought that one might get a clue as to how protein synthesis occurs by looking at the synthesis of actinomycin, a bacterial product that contained only 5 amino acids linked together. The finish line could have been anywhere, and there was lots of speculation about the nature of the code. It did not make sense that every two letters in the DNA coded for one of the 20 amino acids since an alphabet with only 4 letters, in which every word contained 2 letters would only give 16 combinations. In theory it made sense that each code word had to contain at least 3 letters, which, in an alphabet having 4 letters, would provide a minimum of 64 code words, more than enough possible combinations to code for the 20 amino acids. It also was possible that the code words for the 20 amino acids might have varying amounts of letters, but there had to be at least 20 words.

Nirenberg decided to focus his research on protein synthesis, using an extract from simple bacteria, suspecting this approach might be the path to the end of the rainbow. As he wrote, quite confidently, in his lab notebook, "would not have to get polynucleotide synthesis very far to break the coding problem. Could crack life's code!" It was an ambitious statement. Nirenberg and his post-doctoral fellow, Heinrich Matthaie, did a rather simple experiment. To the bacterial extract they added all 20 amino acids but one of them, phenylalanine was radioactively labeled. If the amino acid was incorporated into a protein like structure they could determine that since they could easily separate any radioactive protein that was made from the original radioactive phenylalanine. He also needed a source of genetic information and for this experiment he used a synthetic RNA that was a long chain like RNA, but contained only one of the 4 letters in the RNA, the letter U. This synthetic RNA was called poly U. Early on a Saturday morning in 1961 Matthaie set up the experiment and after an hour incubation measured the amount of radioactivity in the protein fraction. All of the control test tubes reported a background level of 70 counts – while the reaction mixture that contained the poly U showed 38,000 counts per milligram of protein. Other control experiments convinced them that they had successfully made a protein that contained a chain of phenylalanine molecules, one of the 20 amino acids, from the information in the synthetic RNA molecule, poly U. The genetic code had been broken. A series of U molecules in the RNA coded for the amino acid phenylalanine.

This poly U breakthrough experiment was not immediately announced. It was kept secret and only those people working with Nirenberg at the NIH knew what Nirenberg and Matthaie had accomplished. Herb Weissbach, whose lab was two floors below Nirenberg's, first learned about the experiment on the trip from Paris to Moscow to attend the International Congress of Biochemists being held in Moscow in August of 1961. He was on the plane with Paul Berg, who almost two decades later would receive the Nobel Prize, asked, 'Herb, are you aware of the discovery Nirenberg has made that I heard in Paris?' "

Weissbach shook his head and Berg explained Nirenberg's results. "For the first few minutes," Weissbach later recalled, "I didn't fully understand the importance and future ramifications of the poly U experiment. Then the full impact became apparent. I also couldn't realize it at that time, but I just heard Marshal Nirenberg announce my future.

"The first time the scientific world heard about the poly U experiment was at that meeting in Moscow. I was there to give a talk about the work I had done on vitamin B_{12}. When we arrived in Moscow Francis Crick told the organizers they needed to schedule a special symposium at which a major announcement would be made. The auditorium at Moscow University was packed when Nirenberg broke the news that he had broken the genetic code – and then revealed the first code words."

Everyone realized that what they heard was going to revolutionize biomedical research. Ironically, it did not receive much attention in the non-scientific world, because this was the same week the Berlin Wall went up.

That first experiment that cracked the code did not prove how many U molecules in the chain were the code word for the amino acid phenylalanine. The race was on to identify all of the 64 genetic code words. Meanwhile, former Nobel Prize recipient Severo Ochoa was making progress in his own lab at NYU Medical School, as was Har Gobind Khorana, who was then at the University of Wisconsin. Medical doctors could satisfy their draft requirement in the 1960's by spending time in the Public Health

Service. One of the many talented physicians who joined the Nirenberg laboratory was Phil Leder, who showed that only 3 U molecules, a triplet, in the RNA were the code word for phenylalanine. By 1965 Nirenberg, in large part because of the work by Leder, led the race in completely deciphering the 3 letter genetic code. Science finally spoke DNA. All of Nirenberg's work had involved bacteria and the last piece of the puzzle was to prove that it was applicable to all species. Nirenberg's experiments with guinea pigs and toads proved the code was almost universally applicable. He later remembered sitting in his office, looking out his window and marveling at the fact that a human being, a squirrel and a tree were so biologically similar.

It should be mentioned that the Nirenberg effort involved more than 20 other NIH scientists, post docs and technicians who put their own work aside to assist in this effort. In fact after the code was initially broken it was clear to Nirenberg that he would need more staff and space if he was going to win the race and decipher the genetic code. The NIAMD was not in a position to provide what he needed and Marshall was seriously considering leaving the NIH.

In 1961 Nirenberg married a scientist from Brazil, Perola Zaltzman, who was working as a technician in Sid Udenfriend's lab. One day, long after her husband had done the poly U experiment, she came to work in tears because he was thinking seriously about accepting a job offer in Michigan. Udenfriend wanted to keep him at the NIH, and perhaps as important, not lose Perola. He called his senior scientists, Herb Weissbach, Gordon Guroff and John Pisano and discussed how this could be done by contracting their laboratory space, and with the help of Bob Berliner, Head of the Intramural program in the National Heart Institute, by obtaining additional space and positions for Nirenberg. Udenfriend managed to get everything that Nirenberg needed and by 1963 he was able to establish his laboratory contiguous to the labs where Herb Weissbach was working. Nirenberg – and Zaltzman – stayed in Bethesda.

It was becoming clear that traditional barriers between science and industry were falling. As Herb Weissbach pointed out, "People had recognized that the discoveries in molecular biology and molecular genetics, would be the driving force for the development of new drugs for decades to come. We were on the threshold of a new era in scientific research and discovery."

The launch of the Russian satellite Sputnik in October, 1957 had shocked and scared the United States out of its scientific lethargy into a new and bold era. In 1961 President John Kennedy announced his intention to put an American on the moon before the end of the decade. Perhaps for the first time Americans fully understood that the future was being conceived in laboratories. Science, all areas of science from the classroom to the lab, suddenly got the full attention of the government. Additional funding became available for the rapidly growing NIH. By the mid-1960's Nirenberg and others had opened up that door to the possibilities of molecular biology. New research instruments, including nuclear magnetic resonance, high-pressure liquid chromatography and the first rudimentary computers, became available and helped speed up the process of discovery. And at the same time the incredible success of librium and valium had proved to the pharmaceutical industry that it was worthwhile to make large investments in research and discovery. While in the past drug companies preferred to offer numerous products, it became obvious that a single 'miracle drug' could be more valuable than shelves of moderately successful products.

This new American initiative could not have come at a better time for Roche, which was flush with valium profits and looking for something to do with that money. In a 1966 speech given at a manager's conference in Basel, Switzerland, Roche Nutley President Dr. V.D. Mattia described the current situation – and laid the groundwork for what was to come -- explaining, "We are adapting our internal organization to future needs. A company that someday will offer compounds that can alter hereditary potential, chemicals to change the sequence of nucleic acids in viruses, and drugs that profoundly modify thought and emotion – such a company cannot do business in the same old way using time-worn methods. Unless we can fight the battles of the laboratories and drawing boards in a new and more imaginative way, we are doomed to mediocrity. In the future there will be little room in the market place for the 'also rans,' the faint of heart and the imitators.

"We in Nutley are adopting new techniques, employing new methods and testing new and better ways of increasing efficiency and improving effectiveness in all our operations... This is indeed a major risk we are taking, but remember progress often means risk. But we believe the price of failure is not nearly as great as the cost of doing nothing... We also know that good long-range planning can substantially enhance our ability to master the winds of change, to share in the surge of progress and to move into the vanguard of the rapid march of scientific research...

"To build a great company in a great society, we must not only act but we must also dream. We should dream of noble goals and high aspirations, of victories over illness and pain, poverty and suffering. For the great achievements in the history of man have been brought about by those who dreamed greatly and then acted upon their dreams."

The question of how to begin transforming those dreams into reality was answered one night in March, 1967, by Sid Udenfriend and John Burns at a reception in Bethesda for the then NIH-Director Dr. James Shannon, hosted by Steve Brodie. Udenfriend was still running his lab at the NIH, but Burns had recently left Burroughs Wellcome, where he had been Head of Research to become Executive Vice President for Research and Development at Roche in Nutley, New Jersey. Burns had been brought to Roche by Dr. Mattia, to make innovative changes in the way that company conducted research. Legend records that during that party, Udenfriend later recalled, "Late in the evening we found ourselves alone in a room with Dr. Burns, who was speaking glowingly about the planned expansion of biological research at Roche.

"I was impressed, but pointed out that an expansion along traditional lines, limited to mission-oriented research and long-range planning, would probably fail to attract those scientists who were at the cutting edge of the new and rapidly growing biology. I suggested that one way to attract top-notch biologists to Roche might be to earmark some of the new biology positions specifically to staff a newly-created non-mission oriented group. The new research entity I had in mind would be related to Roche in the same way that Bell Laboratories is related to AT&T.

No one knows if this was a concept Udenfriend had been developing for some time or if it simply popped into his head at that moment. Those people who knew him best admit either one was quite possible, but Herb Weissbach thinks it was typical of Udenfriend to throw out ideas on the spur of the moment similar to how he used to come up with experiments to do. But it was one of those rare ideas that sparked a fire. It turned out that the timing was absolutely perfect. Mattia had only recently recruited John Burns and made him vice-president for research. His first assignment was to devise a way of enlarging and strengthening Roche's biological research operation.

As Udenfriend conceived it, unlike existing research programs at most pharmaceutical companies this Institute would not be product driven; instead it would be set up more like the NIH, in which scientists received direct funding, a reasonable time commitment and the freedom to pursue a research project of their own choosing.

It was an audacious suggestion; Udenfriend essentially was dreaming about a place where scientists would have complete freedom to pursue their own curiosity without having to worry constantly about funding, timetables or outcome. It would be a place where pure scientific research would be done without any established goals – and without concern about funding. In other words, a scientific Camelot.

Burns listened seriously. He'd been brought to Roche to make changes and this certainly would be a significant change. There actually had been a much smaller program like this at Burroughs in which an NIH contract allowed scientists in one lab great freedom to do pure research. It wasn't referred to as an Institute or in any formal way, but it was a place where scientists were free to follow their curiosity and their instincts. John Burns had seen how well that had worked, and was intrigued by the prospect of creating a similar environment but on a much larger scale.

The obvious argument against such an Institute was that a pharmaceutical company was a profit-oriented business and research needed to be product-driven. No sensible business model allowed millions of dollars to be invested in projects without any specific direction or potential financial return. Somehow Udenfriend had to convince Burns that creating this Institute made good business sense for Roche. As Herb Weissbach later wrote, Udenfriend explained to Burns that "the benefits to the company would come from the cutting edge research that would place the company in a unique position to move rapidly into new areas of emerging biology and develop novel therapeutics. It was becoming clear even by 1967 that discoveries in molecular biology and molecular genetics that Udenfriend was so aware of because of his association with the Nirenberg laboratory, would be the driving force for the development of new drugs for decades to come."

It was a compelling argument. Like almost all pharmaceutical companies, Roche was focused on organic synthesis. They had nothing at all to do with modern biology, they had no real interest in molecular biology or molecular genetics. This concept presented an opportunity for Roche to build a knowledge base in that area that might prove essential in the not-too-distant future.

To Udenfriend's surprise, Burns took his suggestion seriously. "One would have expected that such a gratuitous suggestion would be dismissed as just so much idle cocktail party chatter," he said. "But much to my surprise, within three weeks Dr. Burns called me with the news that he had discussed the concept with Roche President Dr. V.D. Mattia and that Dr. Mattia seemed genuinely interested."

'Barney' Mattia was an unusual man. For example, rather than participating in an industry-wide advertising campaign to improve the somewhat negative image of drug companies, he agreed to provide medications to people who couldn't afford them. Under his direction Roche gave away an estimated 37 million pills annually. "I can't honestly say that drug prices are cheap," he admitted, "I think they are reasonable only for those who can afford to pay the price. People are entitled to purchase drugs at the lowest possible cost."

A new kind of research initiative was exactly the type of project that appealed to him. As he once told a Life magazine reporter, "The most satisfying aspect of the job of president is to exert leverage for good – to take the massive resources of this company and assign them to projects which offer promise and hope therapeutically, to mobilize its talent and resources to help in the community."

Mattia recognized the potential benefits; in addition to providing a window into this new field of molecular biology, it would help create a very positive corporate image within the scientific community. He grabbed hold of the concept, telling Burns "Why don't you organize it." Burns, in turn, told Udenfriend, "It's acceptable to the administration of Hoffmann-La Roche and they would like you to be the first director." There was still a bureaucratic process to go through, the concept – and the money – would have to be approved by management in Basel, Switzerland, but Mattia was confident that would not be a difficult process.

In early April John Burns told Udenfriend that several members of the corporation's top management were due to arrive in Nutley within a week – and he needed rough plans and a proposed budget to present to them. It seemed like an impossible task.

Udenfriend immediately enlisted Herb Weissbach to help him plan this unnamed Institute. By that point they had become close friends and scientific colleagues. Udenfriend essentially had shaped Weissbach's scientific career. "Sid Udenfriend was a broad thinker," Weissbach recalled. "We worked together extremely well; even when I was a graduate student he didn't like details and I could handle details. Working in his lab I very quickly caught on to the business of conducting an experiment: How do you set it up, how do you execute it and how do you interpret the results to set up the next experiment.

"He'd come in every day with a new idea, let's try this experiment. I hadn't finished the experiment he'd given me yesterday when he was discussing what needed to be done tomorrow. After a while I simply decided only to do those experiments I thought were the most important ones and he usually forgot the other experiments he had suggested. Our talents just fit together so well."

The timing of the new Roche Institute was perfect for Herb Weissbach. By 1966 much of the research in his laboratory was already focused on protein synthesis. It was during this period that he met, and collaborated with, several of the scientists in the Nirenberg laboratory. Among them was Sidney Pestka, who was also working on the mechanism of protein synthesis, and who would later be a colleague at Roche.

In 1966 St. Louis University offered Herb Weissbach an attractive position as Head of the Department of Pharmacology and an opportunity to establish a new Institute of Molecular Pharmacology. Monsanto, a company in this the St. Louis area, was very supportive because they could see how molecular biology technology could be important in the plant field. Monsanto accurately foresaw the importance of genetic engineering of crop seeds. This would have been the first Institute of molecular pharmacology -- but it depended on getting funding from the NIH. Weissbach prepared a grant to the NIH based on this new Institute; it was quite detailed and included the size of the proposed building, the basic departmental structure and the size of the staff. Before writing the grant to the NIH, he had contacted several colleagues to help out. He asked his brother Art to write a section on virology and Syd Spector on immuno-pharmacology and if it was funded he intended to invite them and Nat Brot to join him there. The grant request was close to $50 million dollars, which included construction of a new building and years of funding. At that time it may have been one of the single largest requests ever submitted. And then he waited.

In early 1967 the NIH inspected the site and the plans and responded that some funds would be available – but not enough to pay for a building. Weissbach decided that without a new Institute he was not interested in moving to St. Louis to become Chairman of a department. "It was soon afterwards that Udenfriend contacted me," Weissbach remembers. "'Forget about St. Louis,' he told me, 'we're going to do something much better at Roche.'"

Roche was interested in establishing its own Institute, Udenfriend told him. The two men discussed a broad concept of how to proceed. Weissbach suggested that "if Roche was going to do this, it should be an Institute of Molecular Biology, not molecular pharmacology, because I was concerned about having a pharmacology Institute in a pharmaceutical company. Everyone would think the company would be controlling every decision." In response to Burns request, Udenfriend and Weissbach quickly adapted his NIH grant for the St. Louis University Institute to the Roche plan.

Weissbach's proposal became the basis for the presentation. "By using information from these plans and by juggling some imaginary figures, I was able to produce a three-page document detailing the establishment of a Molecular Biology Institute. This called for a multimillion dollar building and a staff of 20 to 30 independent investigators who would be allocated a suitable budget. Udenfriend later recalled, "I considered this proposal to be little more than wishful thinking, but mailed it to Dr. Burns in time for the Monday morning meeting with Basel's management. To my astonishment, Dr. Burns called me shortly after the meeting and said Roche management had approved the three page proposal for an Institute along with a suitable budget.

"The speed of the whole process – a little over six weeks from cocktail party to approval – left me speechless. Even while drawing up the proposal, I myself had never taken the idea seriously. Now, however, the idea was a reality and I accepted the directorship, subject to Roche's ability to recruit a staff of the highest caliber."

In early May, 1967, the concept was officially approved by both the Roche Executive Committee in Nutley, New Jersey and corporate management in Basel, Switzerland. Among the strongest Swiss supporters of the plan was the head of Research there, Alfred Pletscher – who years earlier had worked briefly in Udenfriend's lab at the NIH with Herb Weissbach. Four men, John Burns, Sid Udenfriend, Herb Weissbach and Alfred Pletscher, who had known each other through working in the Brodie laboratory at NIH in the 1950's, had found themselves in positions to do something extraordinary: create a research center at which scientists would be completely supported, enabling them to turn their full attention to any research project of interest to them without the distractions of teaching, grant applications, and restrictions placed upon most scientists in industry. It would have the culture of the NIH.

The plan to build a yet unnamed Institute at which pure scientific research would be done was set in motion – now all that had to be done was everything.

Chapter Three

Building an Institute

Sid Udenfriend once marveled, "The rapidity of the creation of this Camelot of the biomedical sciences, two months from conception to reality, was not comprehensible to those in academia or government." It was possible only because Roche was a privately-owned family company, so the decision to move forward could be made quickly. Irwin Lerner, who would later become president, Chairman and CEO of Roche Nutley believes "If we had been a public company in the United States it would have been very difficult to establish the Institute. How do you sell a vague idea of what the Institute was going to be, and what it was going to cost, to a stockholder who was only interested in a return on his or her investment?"

Creating the proposal for a unique, independent Institute that was presented to Roche in Basel was exciting, but once the corporation decided to move forward quickly, the reality of what had to be done set in. As the founders were quickly to learn, the location was not one of the strongest recruiting tools. "Nutley, New Jersey at that time was essentially an industrial area – and not especially enticing. In fact, Herb Weissbach remembers that "While I was considering making the move I made some trips to the campus. They were trying to recruit me, so they would pick me up at the airport in a limousine – but instead of driving directly north to Nutley we would drive west and reach Roche by a roundabout way. Having lived in New York I knew that area, and asked the driver why he was taking this out-of-the-way route? He told me he had been instructed to take Rt. 22, a scenic route. But after I finally agreed to come to Roche they still picked me up – but drove straight up the New Jersey turnpike through Secaucus, which was not very attractive".

Family issues also arose. Renee Weissbach, for example, was unhappy with the prospect of moving back to the New York area, especially since their four pre-teen children at this time were happy doing well in the highly regarded Montgomery County, Maryland school system. Sid Udenfriend's wife, Shirley was also very much against the move, especially with two teenage children. Udenfriend finally told John Burns that he was having second thoughts about the move.

It wasn't simply the logistics. Weissbach told Burns that both he and Sid were concerned that they might not be able to convince good scientists to move into industry. Burns knew that Udenfriend had to make the move if the Roche Institute was going to move forward, and sensed Sid's reluctance, especially if Weissbach did not make the commitment. Shortly after Basel officially approved the Institute Burns called Weissbach and asked him to help convince Udenfriend to accept. Among other items, he offered Weissbach the position of Associate Director of the new Institute, adding that Roche would guarantee contractually that he would become Director, if and when Sid stepped down.

In response, Weissbach asked Udenfriend point blank, "Sid, do you want to go to Roche"?

"I really do," he said, "but I want you to come too." He offered to make Weissbach his co-director, which Herb did not want.

A career in science carries with it both great risk and reward. At some point most young scientists have to make a decision that will form the basis of their entire career, often having to make huge leaps of faith. Weissbach knew that the decision to go to Roche had high risk. Almost all of the other scientists with whom he discussed the potential move warned him not to go to Roche. 'Whatever they promise you,' they warned, 'in industry promises can be easily broken." In addition, the NIH wanted him to stay. Bob Berliner was willing to promote him to Laboratory Chief in the National Heart Institute if he stayed.

The only colleague at the NIH who urged him to accept the offer was Earl Stadtman, a former student of Barker, whose opinion meant a lot to Herb. Stadtman told him, "Herb, this is a really exciting opportunity for you. You might have a chance to change the complexion of a major pharmaceutical company."

With all of these thoughts buzzing in his head, Herb looked at Sid and, remembering that Udenfriend had been so supportive when he was suffering from the stress while a graduate student, he said simply, "Let's do it. You should be the Director." The journey had truly started.

The initial decisions had to be made very quickly. The two men had to decide who to invite to join them in establishing a new Institute in Nutley, New Jersey. They had to find a suitable administrator to set up an office. A building had to be designed, and temporary space had to be made available in Nutley for the first scientists to move there.

The first hurdle to be overcome was the suspicion from scientists invited to join the Institute was that it was just too good to be true. John Burns became actively involved in the process. "John Burns was a great negotiator," said Richard Snyder, the former Institute administrator. "At that time he was invaluable. If there was something that we wanted that he knew would be difficult to get through the company, he would continue maneuvering until he found a way to get what we wanted."

Herb Weissbach knew John Burns from the time that they were both in Brodie's lab in the 1950's. and had a good relationship with him. Burns and Art Weissbach had also had contact many years earlier, when Weissbach was studying at Columbia. Burns had been invited to give a lecture about vitamin C. "He had a soft voice," Weissbach remembers, "So soft at times it seemed as if he was mumbling. As he began his lecture it was difficult to hear him. Finally one of the Professors had to urge him several times, 'Speak up, Dr. Burns. Speak up!' From that moment he was forever – and affectionately known – as 'Speak up Dr. Burns.' But he was the perfect person for the job facing him at Roche. He was well liked, and respected. There are problem creators and problem solvers; 'Speak up Dr. Burns' was a problem solver."

Burns suggested that the scientists who were considering coming to Roche should meet with Roche President Mattia, and arranged for what turned out to be a critical meeting. The meeting took place in Mattia's conference room in Building 1 on the Roche campus.

A small group of dubious young NIH scientists had the opportunity to discuss their concerns. Among their real concerns was the possibility that they would give up the security of their present positions at the NIH to join the Institute – and then within a few years Roche would change its corporate mind and shut it down. Mattia listened quietly. He then assured the group that Roche would keep its commitment and provide a building and research support. He emphasized that the Institute would not be judged by the products produced, but rather that "in a reasonable period, such as 10 years the Roche Institute would be an internationally known Center of Excellence". He then said flatly, "Don't worry. If I say it's going to be, that will be sufficient." The scientists left that meeting visibly impressed with Mattia – and Roche – but yet there was a lingering doubt.

Shortly afterwards Burns arranged a private meeting between Herb Weissbach and Mattia. "Barney Mattia was a physician who didn't really understand basic science but his head was screwed on right," Weissbach remembers about that meeting. "He admitted to me, 'I don't understand what you guys are doing but I have a gut feeling this is the way to go.' One of the things we discussed was that we wanted to have post-doctoral fellows in the Institute. He didn't know what post-doctoral fellows did. When I explained that they would spend two years working and training at the Institute then leave, he found that hard to believe. 'We can't have people coming here and seeing what's going on at Roche, and paying them and then having them leave for another company.'

"Here it comes, I thought. I knew this wasn't going to work. I was glad that I'd discovered that before I'd made a commitment. But I added, 'In addition, we want to have graduate students here, we want to be able to train Ph.D. students."

This was the antithesis of the way pharmaceutical companies operated. Corporate science was profit oriented. The drug companies hired scientists, directed their work and owned their output. The science was often directed by marketing: If marketing decided the company needed a drug in a certain area scientists would be assigned to find it. These scientists and researchers worked only on those assigned tasks and were not permitted to publish their results without the permission of the company. The company owned everything that took place in their labs. These companies had no interest in training young scientists. It was a result-oriented business. So naturally Weissbach was surprised when Mattia accepted all the conditions. All of them. Based on Mattia's support, trusting Udenfriend completely and feeling secure with both John Burns and Alfred Pletscher as Heads of Research at Roche, Weissbach finally agreed to help establish the Institute.

There was a brief debate about what the name of this Institute would be. But Udenfriend and Weissbach knew that the primary research done there would be at the molecular level, so the name that eventually would be put on the front door was obvious: The Roche Institute of Molecular Biology.

The concern that the Roche Institute would be a truly academic research Institute in the middle of the corporate world proved to be a substantial problem when recruiting began. Concrete evidence was needed, which led to the suggestion that Roche put the entire agreement on paper: Udenfriend and Weissbach asked for an Institute Charter. This charter would clearly define the academic nature of the Roche Institute as well as the philosophy it would be based on. The concept was readily accepted by Burns and Mattia, and Udenfriend and Herb Weissbach discussed the main points in the charter with John Burns. It was agreed that the charter should describe the Institute's objectives and its working relationship with Roche. Those people drafting it truly were working from a blank page. As this was a unique situation there were no precedents on which to base it. The Institute was to be a place where the strengths of corporate science met the possibilities of an academic lab.

Basically, the intent of the charter was to affirm that scientists at the Institute would be free to pursue anything that

113

interested them without any corporate direction or interference. Signed by Dr. Mattia on July 20, 1967, it began, "The Roche Institute of Molecular Biology is dedicated to fundamental research in biochemistry, genetics, biophysics and other areas in the domain of molecular biology. The Institute will be wholly devoted to long-range basic research designed to shed light on the fundamental life processes. "Scientists at the Institute will enjoy independence in their own choice and pursuit of research problems, guided solely by the scientific importance of the project." After describing several operating principles, including criteria for publishing, a promise that the Institute would have its own building on the Roche campus and assurance that Institute members will be permitted to teach at local universities. It concluded with two points unique to a corporate setting, "Leading scientists from throughout the world will be invited to spend a year or two at the Institute, free to pursue projects of unusual scientific importance and "The Institute will also provide an opportunity for promising young scientists to obtain advanced training and to engage in fundamental research under conditions encouraging professional growth and creative effort."

Charter

The Roche Institute of Molecular Biology is dedicated to fundamental research in biochemistry, genetics, biophysics, and other areas in the domain of molecular biology. The Institute will be wholly devoted to long-range basic research designed to shed light on fundamental life processes.

Scientists at the Institute will enjoy independence in their choice and pursuit of research problems, guided by the scientific importance of a project.

The Institute will benefit from the guidance and counsel of a Board of Advisors, consisting of outstanding scientists in leading universities and research centers.

Publications by members of the Institute will be approved by the Director based on the recommendation of an internal board, and will be judged solely on scientific merit.

Members of the Institute will be encouraged to accept appointments at local universities and to participate in university teaching.

The Institute will be located in its own building in an area near the Hoffmann-La Roche research laboratories. This will permit the scientists of the Institute to pursue their individual research programs while having available to them the resources and services of the existing laboratories of the company.

In addition, the proximity of the Institute to the Hoffmann-La Roche research laboratories will create an opportunity for stimulating interchange in an environment conducive to creative effort.

Leading scientists from throughout the world will be invited to spend a year or two at the Institute, free to pursue research projects of unusual scientific importance. The Institute will also provide an opportunity for promising young scientists to obtain advanced training and to engage in fundamental research under conditions encouraging professional growth and creative effort.

July 20, 1967

V. D. Mattia

President and Chief Executive Officer
Hoffmann-La Roche Inc.

The RIMB Charter. Reprinted from Weissbach,H. Reflections on the Roche Institute of Molecular Biology after 20 years. In Research at Roche, Reminiscences and Reflections for Fritz Gerber (1989), editors, Jurgen Drews and Fritz Melchers, Roche Basle.

Missing from this charter was any suggestion of the kind of secrecy so common in corporate labs as well as all the common restrictions normally placed on employees. While not part of the charter, a working arrangement between the pharmaceutical company and the Institute was discussed. Mattia agreed to a completely hands-off policy, meaning the corporation would not interfere with the work being done at the Institute in any way, although everyone agreed it was to their mutual benefit to promote strong professional and social relationships between scientists working in both areas. Eventually that would evolve into everything from a regularly scheduled lecture series to a highly competitive softball league.

Udenfriend insisted that the charter specifically include the phrase "The Institute will be located in its own building." The reason for this was obvious, but it did raise one unexpected issue that proved surprisingly difficult to resolve. While the building was to be constructed on the top of a hill and within very easy walking distance of the other facilities on the Roche campus, Udenfriend and Weissbach wanted to have a separate gate and roadway to provide access to the Institute building around the clock. It wasn't so much the road that management objected to, but rather what they perceived to be a 'private road' represented. Roche management was very sensitive to the potential issues the

Institute would raise inside the company; and whenever possible tried to defuse them long before they became volatile. Corporate management feared that a private road and guard gate would both symbolically and physically set the Institute apart from the rest of the company and foster resentment. Conversely, Udenfriend was convinced the road would make a very clear statement to anyone consider joining the Institute that it was not simply an extension of the pharmaceutical company, but was separate and unique and acted independently. It proved to be a contentious issue but eventually management agreed – reluctantly – to the separate entrance.

An even more difficult issue was the corporation's insistence that the Institute be surrounded by a fence and protected by security. "In those years," remembers Dick Snyder, who became the Institute's first employee when he was hired as the business manager, "there was sort of an anti-military, anti-industry feeling throughout the country. Being a pharmaceutical company Roche was especially sensitive to the dangers and insisted on high security and guards. Udenfriend was firmly against that, pointing out that scientists coming from the NIH and universities would not want to feel they were part of an industrial-military complex. He kept insisting, 'No, no, no. We don't want fences. We want an open, scientific environment.' Eventually we

117

worked out a compromise. Our site was slightly away from the manufacturing area and up the hill, but we did have a fence around us, and our own gate guard."

Perhaps to counter that feeling of exclusivity the eventual plans for the new building did not include its own full-service cafeteria; instead when employees of the Institute wanted hot food they would have to walk down the hill to the cafeteria in the large main building, known as Building 76. Conversely, corporate employees were invited and encouraged to walk up the hill for lectures, seminars and advice, as Institute scientists were available to them.

One issue not clearly discussed in the Charter dealt with intellectual property. The most restrictive aspect of the charter stated that if anything with potential commercial interest should be discovered in the Institute's labs the company would have the first right to try to develop it into a pharmaceutical product. It was clear from the beginning, and no one objected, that Roche would retain the right to own and exploit research that came out of the Institute. The rules were simple: Before a scientist could publish his results in a journal, the manuscript was reviewed by the Institute's Department Head and/or Director. "If we felt Roche was not interested, the paper went out, to the journal" Herb Weissbach explains. "If we felt Roche might be interested we'd send it to the lawyers, who were given at most 48 hours to review the material. Usually they would tell us that the company didn't want to put a patent on this research. But if they felt the company might be interested the paper would be held up for about a week while the lawyers wrote a provisional patent application to protect the finding. After that was done, the scientist was free to publish his or her work. This process rarely took more than a week; and it was considerably better than the rules that applied to Roche's in-house scientists. They often were not permitted to publish their work or had to wait until patents were filed and the company had decided what they intended to do with them – a process that might take years.

With the Charter finally approved Udenfriend accepted the position as Director of the Roche Institute of Molecular Biology in early June 1967; Herb Weissbach received his official offer letter to be Associate Director on June 10th, and immediately accepted. It was agreed Udenfriend would officially start in the summer of 1968 and move to Roche at that time, while Weissbach would close out the laboratory at the NIH and start at Roche in the spring of 1969. These dates were based primarily on when temporary laboratory space would be available on the Roche campus, while the building was going up.

Although Udenfriend had some administrative experience and Weissbach had first thought about running a research Institute in his failed attempt in St. Louis, neither man had a clear idea of how to begin setting up an entire Institute. They had a concept and a promise of funding; all they needed to do was develop an overall philosophy and establish the regulations by which it would be run, build a cutting-edge facility and staff it with scientists and an entire support staff. The type of operation they envisioned didn't exist; they had to figure out how to bring together the corporate and academic worlds. While the NIH was the essential research model, the only similar corporate situation was the famed Bell Labs, AT&T's basic research center. Scientists working there had developed much of the ground breaking technology that has changed the world, including the transistor, the laser and fiber optics. If anything close to a prototype lab for what Udenfriend and Weissbach wanted to build existed, it was the Bell Labs in Edison, New Jersey. As Udenfriend wrote, "The importance of non-mission oriented research in an industrial environment is evident from the record of the Bell Laboratories, the research arm of AT&T. Non-mission oriented research there led rapidly... to new products for AT&T, but they also created new industries... In a similar way basic findings in biological research in Nutley can be rapidly converted into valuable new drugs or diagnostic procedures." The two men visited those labs and maintained contact with executives there until the Roche labs were in full operation.

When the announcement was made public that Roche had committed to funding a research Institute a lot of people were doubtful −many of them inside the company. As Irwin Lerner, at that point the Director of Advertising remembers, "We didn't know anything about it until after the deed was done. And then we didn't know what to make of it. When we finally read the charter describing what it was supposed to be, a lot of us cynics believed it was just a big publicity scam, great for building public image but it would add little real value to the company. It was just too hard to believe that management was going to throw money at an Institute set up not to find drugs, but instead do whatever they wanted to do scientifically. This is madness, we thought."

Not surprisingly, some of the top scientists were equally suspicious. The Head of Chemistry, Arnold Brossi, who eventually was to become a strong supporter of the Institute initially wondered why it was needed, pointing out, "Roche was a successful company that had built its reputation primarily on the strength of its chemical division" and he questioned "why the company should make a major financial commitment to an entirely new field that had no apparent relevance or products in mind."

At the same time Udenfriend and Herb Weissbach knew they needed an administrator to work with Roche management. Udenfriend suggested Richard L. Snyder, a 31-year old manager who had been trained at the NIH. Weissbach was skeptical because Snyder had no background in science, but learned quickly that Dick Snyder was precisely what was needed at that point, a firm manager with the ability to get along with everyone and solve problems. The timing also was perfect for Snyder, as he explained, "I was working in the extramural, or grants program, in the Heart Institute. I had been at this job for a number of years and I wanted to make a change. Philosophically, this fit me perfectly. I felt so strongly that private industry needed to get involved in funding basic research. To be able to be part of that, especially at the very beginning, was very exciting for me."

The fact that he knew so little about science didn't make any difference -- his area of expertise was people; scientists can be high-strung and opinionated at times and Snyder was able to deal with the variety of different personalities being hired. When a massive project is just beginning and countless decisions about budget, personnel, facilities and conditions of employment need to be made, it's imperative that there be someone capable of seeing the whole picture while making sure the I's were dotted. It requires a unique assortment of management skills to get it all done, and Snyder possessed those qualities. While everyone always knew he was in the room, he never got in anyone's way. On July 31, 1967 Richard Snyder, hired as the Institute Administrator, became the first employee to go on Roche salary. He would also serve on the Institute's Executive Committee. When convincing him to join the Institute Udenfriend explained flatly to Snyder, "You would be a partner with me. I'm the scientist, you're the administrator. I don't want to have to do the administration, and you don't do the science."

Udenfriend and Snyder complemented each other. While Snyder handled the nuts and bolts, or in this case the microscopes and test tubes, Udenfriend was, according to Herb Weissbach, "ideally suited to be director of a basic research center serving the pharmaceutical industry. Although the scientists in the Institute would have completely free rein, Udenfriend had the unique ability of seeing a practical application to many of the programs. He was the one-man pipeline. In this way the company always had a direct line to what was happening in the Institute and the opportunity to have technology transferred without interfering with the research philosophy the Institute was built on."

After the charter was signed the job of recruiting top flight scientists began in earnest. As the Institute was going to be set up to resemble the NIH as much as possible, that obviously was the best place to look for the future staff. The Institute was to have 3 Departments; Udenfriend was going to head the Department of Physiological Chemistry and Herb Weissbach was going to lead the Biochemistry Department. The question was what the third department would be and who was best suited to run it. The answer was right in front of them: In his St. Louis grant application Weissbach had included a Department of Virology and that section of the grant request had been written by his brother, Arthur. It made sense that a Department of Microbiology, which would include virology, would be appropriate for the new Institute. Art Weissbach was the obvious choice and Herb assured Udenfriend that he would be overjoyed to have him be part of the Institute, both as Head of the Department of Microbiology and a member of the Executive Committee.

Art Weissbach had been kept informed of what was going on at Roche and had been at the meeting with Mattia. However, there was a little concern as to whether Roche management would have any objections having the Weissbach brothers on the Institute Executive Committee. "Not at all" said John Burns who was glad to see another excellent scientist coming to Roche.

Art Weissbach, had joined NIH in 1953, after receiving his Ph.D. from Columbia University in Biochemistry. As he remembered, "Before accepting a position at NIH I had been warned it was a government lab and thus by definition was incompetent and staffed by a bunch of left-over has beens who couldn't get a really good job in industry. But I decided to take that chance and within a few years NIH had become the preeminent research Institute in the country and one of the best in the world. It was considered an enviable place to be, especially on a somewhat senior level.

"Of course we also had been warned not to go into industry because it was a one-way street to hell. You had to sell your soul and forget about pursuing those things that interested you. So when Herb first came to me and told me about the Roche Institute we were both very hesitant. We both had good positions in a great research center, why would we give that up? We spent considerable time kicking it back and forth wondering if it was worth the risk. We changed our minds several times. A lot of people tried to convince us not to go, but we realized that you don't necessarily make the best decisions by listening to other people's opinions, which are always based on the past, not the future. The world of science was evolving and this seemed like a good opportunity to be part of that change.

"Part of the equation was a cost analysis. While none of us had gone into science for the money – there was no money – Roche was offering to double my salary if I went there. Finally I calculated that if the Institute survived for only five years it would have been worth the financial risk we were taking. We decided it was worth the risk."

Those four men, Udenfriend, Herb and Art Weissbach, and Dick Snyder formed the first Executive Committee of the Roche Institute. Except, of course, there still was not an Institute; it was all just a dream – although the dream now had a charter. "Initially everybody was worried that the Institute was going to be a total catastrophe," Art Weissbach remembers. "We were warned off by Members at the NIH and most of the scientific community. So it wasn't the easiest thing to recruit the people we wanted to join us at Roche. It was very much a pull-push situation. We were trying to pull them in with a big salary and the opportunity to have sufficient laboratory space and people working for them. We emphasized the fact that because this was a start-up they would have a remarkable opportunity to leap ahead with research and advance professionally. Because the Institute was new there wasn't any of the usual politics.

"They were being pushed by the fact that the NIH was being pressured by Congress to become more clinically-oriented. They were investing a lot of government money and the results were not so obvious. So people really were wondering how much longer NIH was going to keep supporting basic research at the level it did in the past. It was not an easy decision for people to leave a job they mostly liked, pack up their families – and a lot of people had young children – and move to Nutley, New Jersey to join an organization that didn't yet exist on the basis of a promise. So we didn't get every person we wanted, but most people saw the possibilities of this unique Institute and grabbed hold."

It took a lot of courage – and in many cases a salary two or three times larger then they were currently receiving – for people to risk their careers for a promise. Most of these people were firmly established and had bright futures – and it was a real gamble to give up that security for a position in industry, in which the facility had not been built.

Udenfriend offered positions to several people working in his lab at the NIH – and two of them turned him down, wondering aloud if this so-called Institute would ever be built.

"We all understood the high risk of going there," Herb Weissbach says. "With few exceptions everyone was telling us that we were making a big mistake. We were faced with leaving the best research Institute most likely in the world at that point for something that was a big question mark. Essentially we were being warned, 'Don't go. If you go it's going to be the end of your scientific career.' But in my case I was ready to make a change and

127

eager to build something new. If I stayed at the NIH the most I had to look forward to was someday being promoted to a GS-15 or super grade. But to accomplish that my life for the next 30 years would mean parking in the same parking spot, going up the same elevator and walking to the same lab every single day."

As Ron Kaback added, "We figured at worst it would last five years and during those five years you'd have carte blanche and whatever the hell you needed."

Eventually the Institute would be staffed by about 25 senior scientists running groups and as many as 200 additional scientists and support personnel. But initially there was a lot of discussion about those people they wanted to try to recruit. "Our initial decision," Herb Weissbach explains, "was that we would bring in younger scientists who already had established reputations and were respected in their fields. They in turn would attract good younger people who wanted to work with them. Then we had to decide what fields we wanted to bring people from. They might not wind up working on those fields, but we wanted to look into the future. For example, I wanted to bring in people working on how proteins are made in the cell. My brother brought in virologists, people who already were involved in nucleic acid, DNA and RNA research, as it was becoming clear that gene expression, was where science was moving."

The Institute's three departments focused on different areas: Sid Udenfriend's Physiological Chemistry department, wanted to do research on how drugs affected cells and how antibodies could be used to assay for drugs and study drug metabolism; Herb Weissbach's Biochemistry department, planned to focus on protein synthesis and membrane biochemistry; and Arthur Weissbach's Microbiology department, would study DNA in human and animal cells using viruses as a probe. Each department was to include six to eight senior people, principle investigators, and each of them would have no more than eight people working for them; this could include technicians, post-docs and graduate students.

The various titles were patterned after universities, which have Full Professors, Associate Professors and Assistant Professors. The Institute would have full members, which were the senior members, associate members and assistant members. Officially they were all Roche employees. The support staff, office cleaning for example, was provided by Roche.

Each of the original members had strong thoughts about the people they wanted working in their labs, but they all had one thing in common: They shared a curiosity about the way the world worked. Asked about the universal traits that make good scientists, Arthur Weissbach replied, "The first one is curiosity. If you're not always asking why or how, you're probably not going to be very good in a lab. The second is persistence, most good scientists live with the problem they are researching 24 hours a day. It's in your mind all the time and you're just not willing to give up until you find an answer. For other scientists it's a 9-to-5 job, they're not willing to make the necessary commitment. The third is enthusiasm; the excitement of discovery and the willingness to do whatever it takes to solve the problem. Curiosity, persistence and enthusiasm; to me those seem to be the common traits that are necessary to be a successful scientist."

The first scientist officially starting as a Roche employee was Nat Brot, who had initially met Herb Weissbach when both men were students in Mazur's class at City College – although Brot was a Brooklyn kid while the Weissbach brothers were from the Bronx. CCNY tuition at that time was less than $10 per semester, so that great college educated a lot of smart, but poor, New York City kids. Brot's father was only 15 when he came to America from Europe, before World War I, and never saw his parents again. He became a butcher in the city. Brot's science career began when his planned engineering career ended, primarily, he admits, because he was a lousy draftsman. He drifted into chemistry, but after graduating from CCNY his academic progress was slowed for five years with medical issues. During this period he worked at Cornell

Medical School and took some graduate courses at Columbia. He then was accepted to Berkeley in 1958 where he obtained his Ph.D. degree.

It was during his first year at Berkeley that he met Herb Weissbach at a lecture; Weissbach was doing his postdoctoral work with Barker. Brot and Weissbach reminisced about the time they had spent in Mazur's class at CCNY. After completing his Ph.D., Brot contacted Weissbach, who had returned to the NIH, about doing postdoctoral work in his lab. He joined Herb Weissbach's lab at NIH in 1963, and after a couple of years there obtained a research position at the NIH. Weissbach was eager to have Nat join him at Roche and asked him to join the planned Institute.

Like everyone else, Brot was reluctant, agonizing over this decision, "It was well known that those few pharmaceutical companies that had tried to do 'basic research' had failed miserably. These small attempts never lasted more than a couple of years. Merck had tried something somewhat similar although I never knew how academically oriented they were as opposed to conducting directed research. I spent 6 years in government service, and leaving that to take a tenuous position in the industrial complex was a scary business. But I trusted Sid and Herb – and I felt that if it failed there were plenty of jobs available.

So I accepted a position as an Assistant Member, which was the Institute equivalent of an Assistant Professor."

Weissbach also asked two postdoctoral fellows in his NIH lab, David Miller and John Hachmann, and his long time technician, Betty Redfield to move to Roche and they all accepted. Sid Udenfriend also got commitments from John Hutton and Donner Denckla.

Art Weissbach wanted to hire Aaron Shatkin, a talented virologist who years later would become the first Member of the Institute to be elected to the National Academy of Sciences for work done primarily in the RIMB labs. Founding members Sydney Spector and Sidney Pestka also were recruited from the NIH. "They were critical hires," Herb Weissbach says. "They eventually demonstrated that our proof of concept could work, the concept being that if you bring in good scientists, provide whatever they need in terms of freedom and equipment and just leave them alone to work, discoveries will be made that will have practical applications. Syd Spector eventually developed a set of diagnostic assays for a variety of drugs and compounds, including morphine, heroin, serotonin and curare that were far more sensitive and simpler to use than anything previously available – and proved to be very profitable products for Roche."

Pestka, who had been in Nirenberg's group at NIH, where Herb Weissbach met him, initially worked on the mechanism of protein synthesis, but soon after moving to Roche started a program to study a natural compound called interferon that was predicted to have both anti-viral and anti-cancer activity. Working with Udenfriend's group, Pestka developed methods to isolate Interferon, which eventually became the third genetically engineered commercial product. His work led directly to Roche forming the lasting relationship with Genentech that would fundamentally change both companies and the biotechnology world. 'The three Sids,' as Herb Weissbach referred to Udenfriend, Spector and Pestka, played critical roles in bringing products to Roche, starting from basic research conducted in the Institute.

The last of the founding Ph.D.'s from the NIH was Ron Kaback, a known expert in the field of membrane biochemistry. Or, as he liked to describe his work, "The lone wolf in the Institute, the only person working on membranes and transport." Weissbach believed this was a field that the Institute should be studying and offered Kaback a position in his Department at Roche. Ron Kaback's explanation of how a scientist works in a laboratory probably summed up the process that was being implemented as well as anyone has done it: "You come up with some idea, do some experiments, and eventually your idea turns out to be total baloney. But along the way, you're taken in a different direction that turns out to be much more important than the idea that originally got you into the whole thing."

Anna Skalka, the sole female founding members, did not come from the NIH. Skalka had been working in genetics under the direction of future Nobel laureate Al Hershey at James Watson's Cold Spring Harbor Laboratory. "At Cold Spring Harbor I was trying to understand the structure of the genome of a small little bacteriophage – a virus that preys on bacteria, specifically *E. coli*- named lambda. It was a very important experimental tool in the brand new field of molecular biology, and eventually would become key to understanding DNA cloning. We knew it was going to be important, but I don't think anybody knew how important.

"At that time there still wasn't a lot of opportunity for women in science. I had run into some nasty chauvinistic experiences in my career. I was hoping to get a permanent position at Cold Spring Harbor. But I remember being shocked when a female faculty member in my department told me after I had given a very good talk that I should not expect an offer from that department because they already had their 'token female.' Right after that I started looking at other opportunities. I had met Arthur Weissbach at different meetings. In late 1967 I was at a conference about advances in DNA replication on Italy's Amalfi

Coast when Art Weissbach asked me if I would be interested in joining this Roche Institute. 'We're trying to do something very different and new there,' he told me. While it seemed to be too good to be true, it actually was nice to be recruited even though I was a woman. I was young and not tied down so I thought why not give it a shot. If it didn't work I could always go someplace else.

"The more I learned about it the more enthusiastic I became. So I accepted the offer to work in Arthur Weissbach's lab." It wasn't quite that simple. Roche required that everyone they hired had to pass a physical. Anna Skalka informed the medical department that she was six months pregnant with her second child and therefore couldn't be X-rayed. Pregnant? Roche had a corporate policy against hiring pregnant women. "I was very indignant when I heard that," she recalls. So was Arthur Weissbach. They got around that rule by giving her a fellowship that lasted until after the birth of her child. After that she passed her physical and was appointed officially.

In all, including researchers and support staff about 17 people from NIH, several of whom had done their undergraduate work at CCNY, agreed to join the Institute. The NIH people wanted to meet to discuss the Institute in Nutley that they would be moving to starting the summer of 1968. Roche agreed to lease an office in Topaz House, an apartment building in Bethesda about one mile from the NIH. Dick Snyder used that apartment as the RIMB Bethesda office for a few months and the group of scientists would meet there in the evenings during the fall of 1967. Scientists would present their studies and as Herb Weissbach remembers "We all got to know each other".

Topaz House, Bethesda, Maryland, the location of the Roche Institute of Molecular biology office during August and October, 1967. Reprinted from Weissbach,H. Reflections on the Roche Institute of Molecular Biology after 20 years. In Research at Roche, Reminiscences and Reflections for Fritz Gerber (1989), editors, Jurgen Drews and Fritz Melchers, Roche Basle.

Snyder finally opened a permanent office at Roche and Weissbach and Udenfriend would fly to Nutley on a regular basis.

Udenfriend dealt mostly with the Roche management and Weissbach was the one man building committee; he would meet regularly with representatives from the architectural firm, Smith, Haines, Lundberg and Waehler. It was clear that the building would not be ready until late in 1970. However, temporary space could be provided earlier. A schedule was set so that the Udenfriend group would move to Nutley in the summer of 1968 and the two other Departments would move in 1969. The concern was that the scientists who had committed to go would lose their enthusiasm during this long waiting period. Obviously there were no facilities available, so Roche did something quite unusual: Starting in 1968 it offered key people paid sabbaticals until the facilities were ready to be occupied. So while Udenfriend, Herb Weissbach and Dick Snyder were talking and meeting regularly, other people were given their full salary to either remain at their current Institution or were permitted to work pretty much anywhere they could find a laboratory. Pay people who weren't working for you? Nobody had ever heard of anything like that. But Udenfriend and Herb Weissbach believed that not only was it necessary to insure these people wouldn't change their mind before the new building in Nutley was ready to be occupied, it also made the important statement that the Institute was going to be unlike anything then in existence,

While they officially were Roche Institute employees, and were being paid by the Institute, they were working around the world. Art Weissbach, for example, joined Francois Gros' laboratory at the Institut de Biologie Physico-Chimique in Paris where he had worked previously with a group of virologists who later would receive the Nobel Prize for research done at the Institut Pasteur on the control of gene regulation in bacteria. Art Weissbach kept in touch with Udenfriend and Herb Weissbach during the period he was in Paris. Sid Spector spent the year learning immunology from world famous immunologist Herman Eisen at Washington University in St. Louis; when Spector told Udenfriend he had decided to focus on this area of research, Udenfriend shrugged and told him, "Go ahead, that's what this Institute is all about, you can do what you want." Aaron Shatkin spent his sabbatical at the Salk Institute in La Jolla, California working in future Nobel recipient Renato Dulbecco. Dulbecco also joined the Institute's first Board of Scientific Advisors. Anna Skalka continued working at Cold Spring Harbor. Sid Pestka, Ron Kaback and Nat Brot all elected to continue working at the NIH for an additional year, although – just like everyone else - their salaries were paid by Roche.

At the same time Udenfriend recruited an eight-person Board of Scientific Advisors that he would describe as "Not world class, but planet class."

Participants in the first meeting of the Board of Scientific Advisors held in October 1967Left to right, Arthur Weissbach, Marshall Nirenberg, Victor Mckusick,; Bernard Horecker, Sidney Udenfriend, John J. Burns, DeWitt Stetten, Herbert Weissbach, Renato Dulbecco, Avram Goldstein, Alton Meister, Alfred Pletscher, Basle, Richard Snyder, Director of Administration, RIMB. Reprinted from Weissbach, H. Reflections on the Roche Institute of Molecular Biology after 20 years. In Research at Roche, Reminiscences and Reflections for Fritz Gerber (1989), editors, Jurgen Drews and Fritz Melchers, Roche Basle

The first meeting occurred in October of 1967 and in this original group of renowned scientists were Marshall Nirenberg, Victor McKusick, who was known as 'the father of clinical medical genetics;' Avram Goldstein, known for being one of the discoverers of endorphins; Bernard Horecker, then the president of the American Society for Biochemistry; virologist Renato Dulbecco, who would be awarded a Nobel Prize in 1975; Alton Meister, Chair of Biochemistry at Cornell Medical School; and DeWitt Stetten Jr., Chair of the Advisory Board, who had recently accepted the position as Dean of the Rutgers Medical School. Also in attendance were John Burns, representing the Roche Nutley executive committee and Alfred Pletscher, Head of Roche Research in Basle, who soon after played a major role in establishing the Basel Institute of Immunology.

Unlike a lot of honorary advisory boards, whose names are used to attract attention and supporters, this was set up to be an active board, to serve both the corporation and the Institute. Initially they would meet with the Institute Executive Committee twice a year to review the research being done at the Institute and suggest changes that might be beneficial or identify problems; then the board would meet independently with Roche management to bring them up to date on everything and anything being done inside the Institute. Management would then provide feedback to the Institute about the issues that the board had brought up. They would also make suggestions to Roche management as to what the Institute might need.

One point about which Udenfriend and Herb Weissbach disagreed was whether the director of the Institute should be an officer of the company. While Udenfriend pushed for it, Weissbach thought it was a clear conflict of interest. He was concerned it would blur the lines, especially as the new Institute was trying to make the point to dubious outsiders that it operated independently. If it was perceived to be an extension of Roche it would make the tenuous relationship with academia more difficult to maintain. "It would look like a conflict of interest," he argued, "if the head of the Institute was also a Vice-President of the company." Udenfriend disagreed, believing it was far more important for the director of the Institute to know what was going on inside the company – as well as develop personal relationships with corporate management. It finally was agreed that the director would also serve as corporate VP and it never presented a problem.

The real fear at the beginning was that simply by its size Roche's large shadow eventually would envelope the Institute and everyone was careful to set up systems that would prevent that. Only the most optimistic people would foresee what actually happened, and how the Institute successfully led the corporation into a new world – and along the way was responsible for adding billions of dollars to that bottom line.

Planning for the building that would house the Institute began soon after the charter was published on July 20th, 1967 and ground was broken within a year. The initial plans were taken from the proposal that Herb Weissbach had prepared for the St. Louis Institute. The total space was acceptable but the design of the building lacked imagination. Within six months the architects submitted a stunning design, which later would receive several industry awards.

The Institute building would be completely separate from the main facility. It would sit atop a hill overlooking the rest of the corporate campus. It would have five functional floors: The entry level was the administrative floor, which included an auditorium which would seat several hundred people for lectures. The second, third and fourth floors would each be devoted to a single department. The fifth floor was to be used for future expansion. While there were elevators, a central staircase led from the first floor to the second, and each science floor had peripheral staircases on the ends. The idea was to encourage people to use the stairs, making it quick and easy to just run up or downstairs when something was needed, theoretically increasing social and scientific interaction between the different departments.

Roche President Mattia was actively involved in the design. As Richard Snyder remembers. "As he looked at designs he would tell engineers and architects, 'This is a unique Institute, I want it to look that way.' And so the eventual design turned out to be unique. It fit perfectly the concept of the Institute as an open facility that encouraged interaction."

"It was a spectacular building," said Barbara Kerr, a long-time administrative assistant. "Of course if you were new to the place you would get lost. But it was set-up to bring people together. In fact, the first time I went up there for an interview I kept walking in big circles. Finally I bumped into this man wearing a white coat who looked like he knew where he was going. So I stopped him and said, 'I'm looking for Dr. Herb Weissbach. I'm here for an interview.'

"He smiled and pointed down the hall. 'Just go down to my office and I'll be right there.' "Every year when the new group of post-docs arrived to begin working there we would laugh about watching them wandering through the halls trying to figure out how to get where they were going." At the groundbreaking ceremony on September 18, 1968 Hoffmann-La Roche President and CEO V.D. Mattia pointed out, "We have deliberately embarked on an entirely new approach to basic research – an approach which combines the freedom of a university environment with the flexibility, the resources and the long range outlook of a forward looking industrial organization... (The Roche Institute) is designed to provide an optimum environment where scientists, like explorers boldly sailing unchartered seas, will delve into the secrets of the individual cell... its gene activity, its enzyme and amino acid metabolism, its biochemistry and regulatory mechanisms of the life process. In a couple of years, approximately 200 scientists will be at work in this building, asking probing questions and attempting to shed light on the very frontiers of molecular biology."

Ground breaking ceremony in September 1968. Left to right, John Burns, V.D. Mattia, Sidney Udenfriend. Reprinted from Weissbach,H. Reflections on the Roche Institute of Molecular Biology after 20 years. In Research at Roche, Reminiscences and Reflections for Fritz Gerber (1989), editors, Jurgen Drews and Fritz Melchers, Roche Basle

While the facility was under construction Roche allotted temporary space for the Institute to begin work in what were called its research and development buildings. Among the first young scientists to join the Institute from the NIH was Art Bolden,

145

who came to Nutley in 1969 "to initiate an important phase of the project – to get the test tubes shaking. My role was to oversee the establishment of functional temporary laboratories in the Department of Cell Biology in a pill-production building. It was an entirely different environment from the plush laboratories at NIH but no one complained about the cramped quarters, shortage of space or the smell of the industrial site... It was an extremely exciting time, creative juices were flowing, magic was in the air."

Preparing the space for utilization was delegated to Richard Snyder, who recalls, "To get the buildings ready we had to modify the existing space without disturbing Hoffmann-La Roche, who would reclaim that space in a year or two. While they had cold rooms, we had to create warm rooms. But the real test came when I had to order equipment for the temporary facilities, things like scintillation counters and centrifuges. We also had to get some very expensive pieces of equipment. That first order must have been for a couple of hundred thousand dollars and soon after I submitted it I got a call from the corporate treasurer. He asked me to come up to his office to justify this order, and when I got there he asked me one question, 'Okay, what is this equipment going to do in a way of giving Hoffmann-La Roche a financial return?'

"That really floored me. I asked him, 'Are you aware of the purpose of the Institute? Why it was founded?'

146

"'No,' he said, 'No one tells me anything.'

"I tried to be very diplomatic, 'Dr. Mattia said that he wants to fully support basic research in the biomedical areas that scientists are going to be free to do their own science. They are going to be allowed to do the research they want without a direction to earn money.'

"I don't think he knew how to respond. That was completely against everything he did every day. A lot of people inside the corporation were very helpful, but there were people who didn't understand the Institute or saw it as a threat. But eventually the budget was approved; I can't think of anything that wasn't eventually approved."

The first temporary space was in Building 58 and was ready for Sid Udenfriend's group in 1968, whereas the space for Herb Weissbach in Building 71, and Art Weissbach in Building 25 was not occupied until 1969. It was not a good substitute: These makeshift laboratories were cramped and under-equipped, making it difficult to do meaningful research. Arthur Weissbach's lab, for example, was in an older building where the company did large scale fermentations. On a somewhat regular basis everyone had to evacuate the building while a shipment of bromine gas was poured into a chute for storage – although no one ever quite figured out its purpose.

As Nat Brot explains, it was pretty much a makeshift situation. "In a regular laboratory researchers would be able to grow the bacteria they needed for experiments reasonably easily. If they needed large amounts of bacteria, for example, they could do that. But we didn't have that ability, since much of the equipment for the new building had not been purchased. At one time, I remember, we were growing large amounts of E.coli and were using an old centrifuge. It was leaking. I had to sit on the floor and hold the pipes together as bacteria flowed into smaller containers, wondering what happened to the state-of-the-art facilities we'd been promised."

The construction of the Institute involved a lot more than brick and mortar. In order to gain credibility within the science world relationships had to be formed that would allow the Institute to attract top-flight young scientists. While Roche management had finally agreed that the Institute would serve as a training ground for students, initially it was very difficult to find a university that would agree to form the same type of relationship that existed between George Washington University and the NIH. The walls between industry and academia were high and strong and colleges did not want their research freedom to even appear to be tainted by an alliance with a pharmaceutical company. The perception was that any scientist working in industry had given up his interest in pure science. "This was a major headache," Herb Weissbach remembers. "Even though we had published our charter and we really were independent, we initially could not get any university appointments. As far as academia was concerned this was just a ploy for the drug companies to get into their labs. This was really troubling; if the Institute did not have the ability to train younger scientists its role as an academic training center would be in doubt. The first place we approached was Rutgers University, which geographically made the most sense for us. In fact, the Dean of their Medical School was the first Chairman of our Scientific Advisory Board. So we were surprised when the faculty of that school turned us down. They did not want to give industry-paid scientists university appointments. The only person who initially agreed to work with us was Abe Mazur from CCNY, who had taught 3 of the 4 Executive Committee members and Nat Brot. Eventually though, microbiologist Sol Spiegelman, who did pioneering work on viruses, paid attention to what we were trying to do. He had accepted an appointment at Columbia to set up its human genetics program. When we explained our problem to him he waved his hand dismissively and said, 'That's a bunch of B.S. If one of your people is doing good science I'll give him an appointment.' We all ended up with appointments at Columbia, and eventually other universities in the New York area. We had broken down that barrier."

Conversely, the Institute also had to establish a relationship with the scientists already working for Roche who initially didn't quite know what to make of a new and wholly separate research group being set up within the company – and being given brand new state-of-the-art facilities. At first there was a bit of suspicion and a lot of doubt; people who had been at Roche for some time and understood the product-oriented corporate culture found it difficult to believe that the company would spend large sums to support an Institute that was not visibly contributing to the bottom line. But for most Roche scientists, molecular biology was still a mystery, so there was very little overlap between the research already being done inside the company, essentially organic chemistry, and the announced purpose of the Institute. In fact, a lot of Roche employees welcomed the new approach, understanding it was necessary to get involved in the new world of molecular biology if the company was to remain competitive in the rapidly changing pharmaceutical industry. And while at first there wasn't a lot of contact between current Roche employees and members of the Institute, there was a collegial respect among people working in the same profession. The Institute early on developed a vibrant seminar program at which Roche scientists were always welcomed and always treated with respect. "That was very important," Irwin Lerner pointed out. "When company scientists attended these lectures it was never with the idea that they were listening to these brilliant scientists, but that everyone was on the same level, that they were all equal and all important."

There was little crossover. Arthur Weissbach did try to bring one Roche scientist into the Institute, a man who had done some good DNA research, but he refused the offer. Years later there would be a bit more but it was never extensive.

The possibility of applying for patents to protect and own the work done by Institute scientists also was a very sensitive issue. The right to patent a discovery was one of the major hurdles separating industry and academia; companies insisted on owning every discovery made in their laboratories while university labs wanted to openly share their research so other scientists could use it in their own work. Roche attorney George Gould was assigned to work with the Institute to file for patent protection for potentially valuable discoveries. Gould clearly understood the problem. "In 1970 a Roche consultant on the faculty at Harvard Medical School made a discovery about peptides that we were interested in patenting," Gould remembers. "After drafting a preliminary application I went up to Harvard Medical School to see if we could get a release. I met the Dean in a grubby old building on the Back

Bay side of Boston. He was polite but firm. It was against their charter to allow any member of their staff to file patent applications, he told me. He said that in 300 years no one at Harvard Medical School had ever filed a patent on any discovery, they had to publish it and make it freely available. That's just the way it was at that time: Industry filed patents, academia shared all research.

"When I started working with the Institute I wasn't sure how they intended to work. Udenfriend was very academically oriented when we met. He had been tasked with attracting world class scientists so he was very sensitive to their needs. Everything revolved around them. Patent attorneys were not his favorite type of people, which was very common among academic scientists. I was warned that nothing from the commercial side should interfere with their ability to do science at the Institute. The concept of academic freedom was foremost. It was very difficult to talk to anyone at the Institute about patenting, they just wanted to do science.

"One of the first papers submitted for publication were the results of work done by Syd Spector, who was still on sabbatical at Washington University in St. Louis. It was a process for making antibodies to morphine, it was a research tool. Nobody was especially interested in it from a commercial point of view, but I had a feeling I'd better have it evaluated. I sent it to a Roche senior scientist. He got all excited by the opportunity to do a diagnostic assay against heroin. We did a quick filing before the publication. It became the basis for the first commercial product to come out of the Institute, and led to the creation of a whole panel of radio-immunoassays which turned into a very successful Roche Diagnostic business called Abuscreen. Equally important it solidified my relationship with Udenfriend, who realized his people could get publication and scientific credit for their work on a real time basis while still giving the company something of commercial value, which would guarantee continuing support for Institute scientists. After that he turned around completely and became a strong advocate of the patent system. He engaged me in the early stages of a lot of research after that. I often gave lectures at the Institute about the patenting process and the importance of keeping good notebooks."

One of the more practical problems that would never be solved to everyone's complete satisfaction was housing, especially for visiting members of the Institute. Nutley, New Jersey was slightly more than 13 miles from Manhattan, a half hour with no traffic, but it was a suburban area with big city prices. The permanent staff was able to afford that and the fact that Nutley was so close to the city made the area even more attractive. In fact,

Nobel Prize recipient Severo Ochoa was living on New York's Upper East Side when Udenfriend recruited him for the Institute in the early 1970s. It was a reverse commute for him, from the city to Nutley. He had a big Cadillac and loved to drive, but he drove like a distracted scientist. Nobody wanted to drive with him but, as Herb Weissbach remembers, there was a time when he was in the car with him. "One gorgeous morning we were on the highway and he was zipping along. The windows were down, and we were talking science and he looked at me and with a big smile and said, 'Isn't it nice they pay us to do this.'"

It wasn't quite as easy for a lot of other people, especially temporary employees who would be there for only a year or two. "It was a continuing problem that we never really solved," Richard Snyder says. "I worked on it a lot. I visited a lot of places, and we rented whatever apartments we could find. Sometimes we ended up paying rent on empty apartments because we knew we eventually would need the space." The Institute rented apartments in nearby complexes but there weren't enough of them available and those they did rent often weren't large enough. As an internal document reported in the mid-1970s, "Housing is the most critical to our ongoing visiting program... We are acutely aware that the greatest potential threats (to the future of the Institute) are not scientific, but social. Even the most brilliant scientist is first a human being, with family needs normally demanding his initial and overriding attention... Stipends can be increased to keep pace with the cost of living... This is an area where the demand for rentals far exceeds the supply. Owners, no matter how preferable they feel Roche people might be as tenants, all like the protection of renting on a first come first serve basis..." Through the years a variety of attempts were made to permanently resolve the problem, but without success.

The Institute was growing slowly while construction of the facility was in progress. In 1970 Roche provided additional lab space enabling Ron Kaback to move his lab from the NIH. From time to time the scientists would walk through the unfinished building and stand in their future lab. This scientific shakedown cruise was going well. While construction was taking longer than expected everything seemed to be progressing – until Roche's 49-year old president, Barney Mattia, died suddenly of a massive heart attack.

Barney Mattia was an unusual man, who believed a company did best by doing good. In 1969 he had been honored by the B'nai B'rith as the International Humanitarian of the year, and was publicly active in trying to prevent the Vietnam War from extending into Cambodia.

Mattia believed creating the Institute would enhance the company's image in the industry – but it was an expensive commitment without a financial upside and there some doubt his successor would feel the same way. His death sent a shockwave throughout the entire Institute. A lot of lives had been uprooted, houses had been purchased and apartment leases had been signed, the Institute building was only months away from completion -- and the biggest corporate supporter of the concept, the man who had convinced the Swiss to make the multi-million dollar commitment was suddenly gone. A lot of people were holding their breaths as management selected a new president and CEO.

Robert B. Clarke was selected to replace Mattia. Bob Clarke had earned his graduate degree in chemistry from Harvard as well as a law degree from Brooklyn College – and had chosen to become a very successful patent attorney. That was not necessarily a good sign for an Institute founded on the principle that patents would be a by-product, not an objective.

Bob Clark had a big personality – and he understood his industry. To everyone's relief he embraced the concept of the Institute and continued to support it without any interruption, while maintaining Mattia's corporate hands-off policy. In recognition of Mattia's contributions to Roche, Clark established the V.D. Mattia

Lectureship, which included a $5000 stipend to be awarded to a leading scientist. The first person to receive the award was future Nobel Prize recipient Paul Berg, who was cited for "fundamental discoveries on DNA, especially on recombinant DNA."

First Mattia Lecture in 1972. Left to right, Robert Clark, Paul Berg and Sidney Udenfriend. Reprinted from Weissbach,H. Reflections on the Roche Institute of Molecular Biology after 20 years. In Research at Roche, Reminiscences and Reflections for Fritz

Gerber (1989), editors, Jurgen Drews and Fritz Melchers, Roche Basle.

The move into the new building began in May, 1971 and within a month the entire Institute staff of 163 people was at work on the three department floors. The top floor initially was left empty. A few of these people had come to the United States under the State Department's Exchange-Visa-Program, which had been established to encourage foreign scientists to work in America – making Roche the only industrial participant in that program. At the dedication in September, Bob Clarke admitted that in return for the large investment in fundamental research, "We hope for products." But in his speech he emphasized his belief that the company understood its social responsibilities if and when "advances in fields such as genetics, antiviral agents, and other areas provocative of social controversy occurred."

In its story about the dedication the journal Science wrote, "The belief that entirely new forms of marketable therapy... may emerge from basic molecular biology lies at the heart of the Roche approach to basic research."

The state-of-the-art labs had been carefully designed to allow expansion into the new worlds of molecular biology that were just beginning to blossom. The once simple laboratories built in the basements or unused rooms of wealthy men to pursue their curiosity had been transformed into expensive high-tech environments in which cutting-edge investigations of micro-science could be done. In many ways these were the labs of the future. In addition to extensive bench space, storage space, shelves and cabinets, the building included several sterile transfer rooms and sterile hood rooms. The labs were equipped with several millions of dollars of the best.   available scientific apparatus and equipment, much of which had been purchased from Beckman. This included an estimated 20 ultracentrifuges, at least as many radioactive counters, numerous high-powered microscopes and a seemingly endless supply of containers, beakers, and everything else that might be needed. In fact so much Beckman equipment was installed that for the first time in that company's long history it assigned a technical rep to work full-time in the building.

As Science magazine giddily described the building, "The Institute differs from today's university department in one significant way – affluence. Though the new building is not equipped with spigots for hot and cold running radionucleotides, there is obviously no need for a researcher to scrounge, either for equipment or materials. As one of the staff put it, "If I needed an ultracentrifuge for my private use, I would have to justify it, but there would be no problem."

The building wasn't simply a physical success. It had been laid out to promote collegial relationships among people working on the different floors and, from the day it opened, that proved to be a hallmark of the building. As Tim Coleman, who joined the Institute years later, explains, "It didn't matter what floor you went to; if you took the time to introduce yourself and talk about what your challenge was people were incredibly welcoming. It was an incredibly collaborative atmosphere. At one point, for example, I had a problem in understanding how you carry out protein translation and Aaron Shatkin, who could write a book on this process, was working on the fourth floor and was always available to anyone who needed help.

"I was using the restroom my first week there and Severo Ochoa walked in. I was stunned as I realized that a Nobel laureate used the same restroom I did. Imagine being a 21-year old kid and sharing a restroom with the man who helped figure out how RNA turned on proteins.

"The doors were never locked. Nobody was hiding secrets. Everyone in the building was accessible. Once, I remember, I had an experiment I had to do on a Saturday morning so I had to get into the lab very early. One of the few cars in the parking lot belonged to John Kopchick, who was a postdoctoral fellow in Dennis Stacey's lab. After I finished my work I went up to the 4th floor to see him. When I asked him what he was doing there that early he told me he was doing some microinjection. That's a technique in which you take DNA from one cell and stick it into another cell to change that cell's function. I had never done it and wasn't sure how to do it correctly. It's a nice skill to have".

"He started showing me how to do it, then asked if I wanted to try. 'Sure,' I said. The next thing I knew I was doing it. Obviously my work wasn't an important part of his experiment, but to have the opportunity on a Saturday morning to learn from a great scientist was just awesome.

"That's what it was like there and I just assumed that's the way every place ran. I found out later in my career that wasn't true. But at the Institute you couldn't go up or down a staircase without bumping into a colleague and talking about what you both were doing. That was the spirit of the building."

While there had been great doubt throughout the pharmaceutical industry that Roche was really serious about funding a pure research Institute, the man who had convinced the corporate board in Basel that this was a worthwhile idea, Director of Research Alfred Pletscher, believed in this concept so strongly that in 1969 he established a second Institute, the Basel Institute of Immunology. Like the facility in Nutley, a primary purpose of the Swiss Institute was to help bridge the intellectual gap between industry and academia. The Immunology Institute was not on the Roche campus in Basel, like the Institute in Nutley but instead was a brief drive away. The Immunology Institute also moved into its new facility in 1971, and similar to RIMB, its two floors were connected by a spiral staircase – to promote the collegial environment and create what Tim Coleman called, the spirit of the building.

"The Institute building was kind of circular," Amiya Banerjee explains. "It was unique. I have visited many universities all over the world. There is always a lack of interaction due to the bigness of the building. You had to walk quite a bit to the next colleague. But at the Institute if you came out of the elevator and made a right turn, within a few minutes you would return to the elevator – and during that time you would encounter several people. It was almost impossible not to run into people. All day long everyone would meet everyone; we would interact with each in the corridors, in the rest rooms, in the elevators, and during that time we would constantly be talking about the same thing: Science. There were blackboards with chalk in the corridor to encourage people to stand there and talk. It was science, always science, everything was science.

"It was not at all unusual to see a group of people flocking together excitedly. When someone asked what was going on the answer often was, 'He got fantastic results.' And everybody would join the group. We all knew what everybody else – not just on our floor, within other floors – was working on.

Chapter Four

The Brave New World Begins

Like Dorothy opening the door of her crashed house to discover the colorful and exciting world of Oz, the Roche Institute opened its own doors onto the possibilities of a brave new world. Aldous Huxley's classic 1931 novel, Brave New World, envisioned a world in which science had made it possible to genetically-design human beings to fill the needs of society. In Huxley's dark vision people were bred in test tubes and given only the intelligence and skills they would require to fulfill their planned future. Families had ceased to exist and individualism of any kind was greatly discouraged.

And due to the incredible advances being made in molecular biology Huxley's fiction was on the verge of becoming scientifically possible. But most scientists didn't envision the same outcome as he did. Instead, they were dreaming about the possibility of eliminating genetically-inherited diseases, curing cancer and changing a host of other potentially deadly biological conditions.

165

The key to being able to alter those natural processes was learning how to manipulate the human cell. As Arthur Weissbach explained, "Once we had learned everything, or at least a lot more than we already knew, about DNA, RNA and proteins, we could start manipulating genes. We were on the right path. We knew that DNA contained the information to make at least 20,000 proteins. We knew RNA carried the genetic message within the cell. We were close. And once we figured it all out we could start moving genes around. And when we did, everything was going to change."

Scientists had learned more about life and death on the molecular level in the previous two decades than in all of history, but it really was the continuation of a chain that had begun in 1590, when Dutch spectacle makers Hans Janssen and his son Zaccharias discovered that if they put several lenses in a tube an object could be greatly enlarged. It was this compound microscope that served as the basis of the development of microscopy done by Galileo and Leeuwenhoek – and led step-by-step over centuries to the discovery of Crick and Watson.

The staff of the Roche Institute had literally been there at the beginning. Ron Kaback, for example, was a freshman at Haverford College when Watson and Crick had published their world changing experiment "Professors began introducing students to this new world," he explained. "There were no textbooks. We used existing literature and Scientific American. The whole thing was taught experimentally – we were being taught it as it was being discovered. It wasn't history, it was current science.

"Being there at the beginning was very cool. I consider myself very lucky in the sense that I could see the whole field as it developed. What really got me was this whole notion of creating something yourself that would be both interesting and intellectually challenging."

The once esoteric world of molecular biology was suddenly the center of scientific research, and the Roche Institute had been designed to advance that knowledge. It also had been built with the hope that a dazzling variety of new drugs capable of treating an array of illnesses and conditions – and the potential for billion dollar profits – lay at the end of that rainbow. The days of experimental science being a rich man's passion had long ago faded into history. In the early days of biochemistry, for example, as Arthur and Herb Weissbach were beginning their careers, Art remembers, "By the time we were starting the Institute I knew almost every major biochemist in the United States. I'd either met them, knew someone who had worked with them, or knew about their work. It wasn't that difficult because when Crick and Watson made their discovery there were less than two thousand trained

167

biochemists in the country." But that number grew rapidly into the 1970's when the lines between the corporate world and academia began blurring, and then exploded when the brand new biotech industry made it possible for men and women to earn fortunes in the laboratory.

The direction that the Institute would take was clear from the first annual report, published in 1969, in which the three labs reported their areas of investigation: In Herb Weissbach's Department of Biochemistry researchers were at work on protein synthesis, cell metabolism and mechanism and regulation of enzyme action. The emphasis in Arthur Weissbach's Department of Microbiology included animal viruses, bacteriophage and cell studies. And Sid Udenfriend's scientists in the Department of Physiological Chemistry were trying to understand the biochemistry of collagen, as well as investigating the biosynthesis of chromatophores, mammalian genetics and neuroscience.

Some of the individual projects in1969 show the great diversity of areas under investigation. projects were under investigation; Nat Brot and Herb Weissbach reported on the Role of Transfer Factor G, a protein involved in the movement of the polypeptide chain on the ribosome. Researchers Brian McAuslan and Kong Beng Ton reported on The Regulation of Gene Expression in Virus-Infected Cells. Arthur Weissbach, Abner

Schlabach and Arthur Bolden reported on the Control of DNA Synthesis in Animal Cells. John Hutton noted that "Information about the chromosomal location of genes with known biochemical functions has been extremely difficult to obtain in mammals," then made strides in doing that and published a study Linkage Relationships of Genes with Biochemical Functions in the Inbred Mouse. In his paper Antibodies Directed Against Morphine and Norpinephrine, Sydney Spector demonstrated that "Since antibodies can be formed in response to small molecules (haptens) when they are coupled to proteins, the technique has great potential in the study of physiologically active compounds," which led eventually to his creation of a diagnostic, and very profitable, assays for drugs of abuse. While still working at the NIH Sidney Pestka, published a paper with Robert Gallo entitled Control Mechanisms in Normal and Cancer Cells.

None of the 50 projects described or the 65 papers published in the first year of the Institute's existence was of itself revolutionary or important enough to change the world of science; they didn't conduct a voice over a wire or make a metal filament glow. Instead they were the same type of small steps that could be built upon. These experiments, reported openly for the use of other scientists, were small steps on a very long road. As Herb Weissbach said, "Success is the result of a series of very small, or what appears to be, very small goals. A scientist doesn't necessarily set out to cure cancer, instead he or she wants to find out basic differences between normal and cancer cells. When people of my generation chose science as a career nobody was focused on the practical application of their research. It was more of a detective story: It begins with an observation and then you ask questions. There are those who use a whole cell approach, and are not primarily interested in the molecular events within a cell. Then there were people like myself, and many others at the Institute, who were trying to break open the cell and answer those questions by dissecting out a small part of the process in hopes that eventually we will be able to put everything together and answer basic questions in biology. And every time we answered one of those small questions it raised a dozen more questions that needed to be considered.

"You need infinite patience, but if you believe your project is important enough then you have to stick with it."

The first substantial commercial success to come from the Institute was Sydney Spector's assay for drugs of abuse. The work he did during his sabbatical with Herman Eisen at St. Louis' Washington University had led him to change his direction. As he described it, "I started to develop antibodies to various drugs to follow their metabolism... I decided to go on a fishing expedition and asked whether one could use antibodies as surrogate receptors, something like a fishing hook."

The concept was simple: He would use specific antibodies to neurotransmitters and drugs, which are highly sensitive, as a means to search for the presence of important compounds in the body. Initially this included antibodies to barbiturates, morphine, reserpine, imipramine, the antidepressant desmethylimipramine, chlorpromazine and haloperidol, in addition to neurotransmitters including serotonin and acetycholine. These tools actually proved to be more sensitive for detecting the presence of these compounds than anything previously available – and they also served as the first indication of how valuable the Institute could be to the company.

The timing was perfect. A year after the 1967 founding of the Roche Institute, Roche entered the world of pharmaceutical diagnostics by purchasing the diagnostics unit of Chemische Fabrik Schweizerhalle. Also independently the Basel Institute of Immunology was founded in 1969 and opened its doors in 1971. A few years after that The Roche Research Institute of Marine Pharmacology was founded in the Sydney, Australia suburb of Dee Why to search for drugs from marine life – drugs from the sea. Spector's discovery intrigued corporate scientists and led to several patents, helping form the basis of what was to become the

Diagnostics Division. It also helped to allay questions of how well – or if at all – Institute scientists could work with Roche corporate researchers.

There could not have been a more fortuitous time for the Roche Institute to have been founded. At the building dedication ceremony Dr. Renato Dulbecco, who was soon to be awarded a Nobel Prize, pointed to the past – and the future, saying, "One of the fundamental developments, in the last twenty years or so, is that biology has become molecular... Investigations have shifted from whole organisms or their cells to the components of the cells themselves, in what are called cell-free systems. A cell free system contains components extracted from cells, which after extensive purification from other components still perform a specific function in the test tube. The biological operation thus isolated out of the complex system of the cell can be investigated without interference from other operations, to identify the molecule that performs it. Since they are the indivisible basis of biological functions, these molecules are the real essence of biology."

It was, as Herb Weissbach, called it the beginning of biotechnology. "We were beginning to develop the tools that would enable us to engineer DNA so that a protein could be expressed in another organism best suited to make it." That would open up a new world, a world in which existing living organisms could be genetically altered and organisms with a new genetic make-up could be created. The potential to change the world was breathtaking; genetic disease could be eliminated, cures for other disease – like cancer – could be produced in a test tube, new drugs to treat almost any disease or condition could be created, and at the far extreme, the secrets of life itself might someday be uncovered. For the pharmaceutical industry, biotechnology promised to be the crossroads of science and commerce. It was the future, the potentially very profitable future.

The foundation had been laid; it was now known that DNA was a molecule consisting of two intertwined strands that carried the genetic code and that DNA directed the synthesis of RNA, which carried the genetic information within the cells to where specific proteins could be produced -- The proteins that determine every individual characteristic; from blue eyes or brown eyes to the color of your skin. Other world-changing discoveries began occurring very rapidly.

Scientists around the world were attempting to take the next small steps; transform this basic understanding of the biological process into a system that would literally enable scientists to manipulate nature.

One of the first significant breakthroughs in the recombinant DNA field was announced in 1971, when Stanford's Paul Berg successfully spliced together strands of DNA from different organisms; in effect, creating, in principle, a new form of life. For many scientists this was an extraordinary thrilling achievement and a necessary step forward in understanding how we might combat genetic diseases and defects, but for others it was the chilling first step towards the creation of a Frankenstein monster. As Renato Dulbecco explained, "Some scientists look at the synthesis of genes and their possible uses in man as the beginning of a new form of the manipulation of Nature, defined as genetic engineering, which reaches into the most intimate machinery of life. Many people, however, are afraid that genetic engineering will have dire future consequences, similar to the exploration of atomic energy for creating means of destruction, and strong argue that the whole field should be left aside." Dulbecco went on to suggest that this technology had extraordinary promise, including the ability to someday treat congenital gene defects – and the diseases they cause – by creating artificial and healthy genes in the laboratory.

Paul Berg was part of the post-World War II generation of young scientists who seem to have emerged from similar backgrounds, who crossed paths with each other in labs around the country and eventually changed the world. It was a small, highly dedicated group, connected by friends and associates, mentors and interests. Almost none of them were ever further away than a phone call, and there was a growing sense among them that they were in the midst of a scientific revolution. They were the first people to explore this new world.

Paul Berg, for example, was another New York City kid who fell into science. He was born in Brooklyn, his father was in the garment industry. As he explained when he was awarded the Nobel Prize in 1980, it was at Abraham Lincoln High that he learned the basic attraction that science held for so many people, "The satisfaction derived from solving a problem with an experiment was a very heady experience, almost addicting. Looking back, I realize that nurturing curiosity and the instinct to seek solutions are perhaps the most important contributions education can make. With time, many of the facts I learned were forgotten but I never lost the excitement of discovery."

He eventually enrolled in City College to study chemistry, before he really could get started he enlisted in the Navy. After the war he graduated from Penn State and got his Ph.D. from Cleveland's Western Reserve University. While few people knew much about Western Reserve, it had one of the best biochemistry departments in the country. By the time he graduated, "I was hooked on a career in academic research instead of one in the pharmaceutical industry. After working at a lab in Denmark for a year he joined Arthur Kornberg's lab at Washington University in St. Louis. He stayed in St. Louis for six years, as he wrote, doing what so many young scientists were doing, "making the slow transition from classical biochemistry to molecular biology and becoming increasingly preoccupied with how genes act and how proteins are made."

In 1971 Berg successful spliced together a minute amount of DNA from a bacterial virus known as lambda with a DNA sample of a virus found in monkeys. This was the first manmade DNA, which became known as recombinant DNA – and resulted in Berg winning his Nobel Prize.

It was an extraordinarily important step forward as Herb Weissbach explains, "Theoretically we now had the ability to cut a section of DNA that contained the information for one specific protein, one gene. The human cell has 20,000 genes, so theoretically we could cut and paste and make essentially 20,000 human proteins in any cell, even a bacterial cell."

While Berg proved it was possible to transplant recombinant DNA from one organism to another he didn't take the step to do it. As with many great scientific advances there was tremendous fear of the unknown. The ability to alter forms of life offered amazing possibilities – and not all of them were positive. Scientists worried that harmful recombinant DNA, perhaps a gene that created cancerous tumors, for example, might escape the laboratory in common bacteria and begin to spread unchecked. There also were vitally important ethical, legal and environmental questions to be answered. Paul Berg became a leader among scientists trying to map a safe path for moving forward.

The next, cautious step forward was taken a year later by Herbert Boyer at the University of California at San Francisco working in collaboration with Stanley Cohen of Stanford University. The two men had met at a conference in Hawaii and both of them realized immediately how well their independent research fit together. Sitting in a deli late one night, they mapped out the plan that led to them successfully splicing together two strands of DNA, inserting it into E. coli bacteria which reproduced countless exact copies of the inserted DNA, or as these copies became popularly known, clones. This marked the beginning of recombinant DNA technology. The world of biotechnology was born.

While this new world was being born the Roche Institute was successfully coming into being. The Institute finally moved into its own building in May, 1971. Perhaps not coincidently, this took place at just about the same time the second pure-research Institute founded by Roche, The Basel Institute for Immunology in Switzerland, also began making progress. That entity, founded by the Director of Research, Albert Pletscher, was established based on precisely the same principles as the Nutley Institute; according to its director, Nobel Prize recipient Niels Kaj Jerne, "The choice of research projects to be undertaken at the Institute must in no way be influenced by commercial motives... Two main considerations have led Roche to create this Institute: First, Roche feels that its significant position in the fields of medical and biological sciences carries with it the obligation to further basic research. Second, Roche hopes that the presence of a highly qualified group of scientists dedicated to fundamental research and working in academic freedom may provide new knowledge and valuable stimulation to industrial research."

By the end of 1971, its first year in the new building in Nutley, the Roche Institute had grown to 163 people, including 33 members, 49 postdoctoral fellows and 37 research assistants. The university connect had been secured with ten staff members holding formal appointments while others participated in teaching programs. In each Department of the Institute journal clubs, where scientists discussed recent published articles, had been established to enable the entire staff to keep up to date with current discoveries in fields that were growing in importance, including gene expression, genetics, neurobiology, virology and immunology.

The biggest news was, of course, recombinant DNA. The entire pharmaceutical industry had immediately understood the potential commercial value of this biological technology. The tall and strong walls between industry and academia began weakening. In California Stanley Cohen decided to stay in the academic world, and eventually ended up being the defender of recombinant DNA technology against those fearful people who wanted to control scientific research, while Herbert Boyer took the other path, joining with venture capitalist Robert Swanson to open a brand new commercial research firm to exploit the possibilities of recombinant DNA, which they named Genentech.

Glancing through a summary of the projects being pursued by scientists at the Roche Institute in 1971 accurately reflect the excitement and curiosity about the new world that was being opened up by men like Berg, and Cohen and Boyer. In Herb Weissbach's Department of Biochemistry research was focused mainly on the mechanism of protein synthesis, membranes and Sid Pestka's group was beginning their studies on interferon. Arthur Weissbach's Department of Cell Biology was concentrating on the way viruses interact with basic cell processes and, as Art

Weissbach wrote, "the mechanism by which viruses express their genetic information... in an attempt to understand the fundamental processes of viral multiplication in cells." And Sid Udenfriend's Department of Physiological Chemistry was studying collagen biosynthesis, neuroscience and immunopharmacology. Syd Spector, one of the primary researchers in Udenfriend's department, had immersed himself in this new area of immunopharmacology. 'New methodology moves science!" he used to say, boldly. Even in those early years he was attempting to develop new tools to measure drug levels in a quantitative way in plasma, brain tissue, and cerebrospinal fluid. 'The Spector Monoclonal Antibodies,' as they were to become known' were making it possible to determine in tissues of barbiturates, morphine, reserpine, imipramine, desmethylimipramine (DMI), chlorpromazine, and haloperidol.

As he later explained, he had learned how to approach research projects from Udenfriend, who told him, "If you're going to look for something make sure the detection method, or assay, is a simple one. You don't want an elaborate method... He helped me find a way of generating studies so that all variables are controlled and all factors are understood." Spector was on the track of something that would prove to be extremely important for both the Institute and the larger company, even if he didn't yet know what it was, and simple step by slow step he was pursuing a concept.

The existence of this new Institute in New Jersey was beginning to attract considerable attention within the scientific community. Silvio Spadari, then a young scientist in Pavia, Italy, remembers reading that "A. Weissbach identified in human cells a low molecular weight DNA polymerase much less abundant than the high MW enzyme present in proliferating cells and in 1972 he also described a third DNA polymerase. While these polymerases were not yet accepted as distinct DNA pols by the scientific community, I was fascinated by the pioneering work, feeling that a completely uncharted territory was just within reach, waiting to be discovered by those bold enough to take up the challenge. I don't know whether it was courage or just naivety, but I was surely eager to be part of the pioneers... In those early years molecular biology was still in its infancy and it was not easy to find, not only in Italy but also in Europe, such high profile research centers..."

The Institute continued to grow. A year later, in 1972, the staff had grown by another 30 people, including 53 post-doctorate fellows. Equally important to the overall mission, which included training scientists, 10 postdoctoral fellows left to accept positions as faculty or staff members at universities and research centers. A culture of excellence was developing. In addition, journal clubs and research lectures were constantly springing up to allow scientists to add to their knowledge beyond their daily activities. These groups mirrored the advances in the growing world of biomedical science. Short instructional courses were given in lab techniques like electron microscopy, fluorescence analysis and rapid reading. Members regularly presented seminars on their own work and journal clubs became an intrinsic part of the Institute. There was a lecture almost every day and a large number of respected scientists would visit the Institute and lecture about what was happening in their fields. It gave the members an opportunity to meet people outside their own area of expertise, people who they otherwise would not interact with, and enabled them to form both social and professional relationships. In those early years there was a recognition that the Roche Institute was in fact a very special place and those people on the staff had been carefully chosen and hired to be able to take advantage of what it offered. All of these activities were supported entirely by Roche.

"There were Nobel laureates and some of the top scientists in the country continually coming through to give a lecture or work there on a sabbatical," George Gould, the Institutes Senior Patent Attorney remembers proudly. "It was a place where young post-docs really had the opportunity to rub shoulders with the elite scientists in their field. And there was a constant technology transfer, just by being in the same building. There was a daily and natural collaboration on the research level. All the barriers had been taken down."

"Severo Ochoa was the first Noble Laureate to join the Institute," remembers Amiya Banerjee, one of RIMB's first post-doctoral researchers. "He would come to my seminar and I would be giving my lecture and suddenly I would see Ochoa sitting in front of the room with his notepad opened. That was so inspiring. And Bernie Horecker, a member of the National Academy of Science who had been my boss as Chairman of the Molecular Biology Department at Albert Einstein College, later became my colleague at the Roche Institute. He would come to my presentations and ask me questions about my work and I would think, 'My God, Bernie is asking me scientific questions. There were no bars between some of the best scientists in the world and the rest of us trying to learn how to be scientist."

In April, 1972, the Institute began running the symposia on selected topics that would continue for two decades. 50 scientists from the United States and abroad attended the very first symposium, "The Biochemistry and Metabolism of the Arene Oxides." As hoped, it proved so successful that the second symposium, Genetic Recombination, was scheduled for the following year.

The Institutes first V.D. Mattia Lecture was awarded to Paul Berg in 1972. The Mattia lecture eventually became one of the most important annual events of the year. As Tim Coleman remembers, "There was a seminar series on each floor once a week and journal clubs that met every day. In addition to people working in the building speaking to the other departments, people from outside were invited in to give presentations on their work and how we might collaborate.

"But the Mattia Lectureship was the highlight of the year. At least eight of the recipients eventually went on to win a Nobel Prize. It was a grand affair. It took over much of the building. The lecture had to be piped into a large reception area outside the lecture hall so people who couldn't get into the over-flowing lecture hall could hear it."

The existence of the Institute continued to gain recognition throughout the scientific world, although there was still considerable doubt that it would remain free of pressure from the industrial side to work on specific projects that might result in commercial products. But the opening of the building had made an initial splash. While research labs throughout the country were struggling for survival – in many places supposedly disposal test tubes were being washed and reused to save money – the fully-funded Institute stood out. The new building was so well-equipped that Beckman, a premier manufacturer of laboratory instruments at that time, such as radioactive scintillation counters and centrifuges, actually had a technician working full-time on-site to deal with any mechanical problems. As Joe Tarnowski, a post-doc who joined the Institute in the late 1970s said, "If you couldn't do good science there, you couldn't do it anywhere."

In fact, by the end of 1972 the Institute had gained such incredible respect that Exxon, one of the largest energy companies in the world, wanted to establish a collaborative relationship. At that time the oil companies were as flush with cash as the pharmaceuticals. There was a big push to use those profits to prepare for the future, to figure out how to move from simply pumping oil out of the ground into establishing more technologically advanced production. They were just beginning to move into biofuels, for example, an area in which they had absolutely no expertise or had even established a viable research capability. Senior patent attorney George Gould recognized the potential for a profitable relationship. "It seemed like a natural fit," he said, looking back. "Inside the Institute there was always some concern about our future. It was sort of a, this is too good to be true, anxiety. Roche had spent a considerable amount of money putting up a magnificent building, they were supporting a lot of labs and a large staff. There was some grumbling from inside about it and people wondered how long Roche was going to be able to afford paying for all of this. Personally, I always felt a drumbeat underneath the surface questioning the viability of the Institute. It seemed to me that the leaders of the Institute were always concerned that somebody somewhere in corporate research was going to get some real power and second guess the concept of siphoning off funds from the corporate research budgets to do basic research. I know that when the budgets were announced every year there was a sigh of relief inside the Institute.

"I thought we could off-set some of the costs by bringing in a partner with deep pockets. That was Exxon. I had some contacts at Exxon and I floated a proposal, suggesting they might be interested in doing a collaboration. They could send their people there to be trained or have a first look at any inventions coming out of the Institute that might have applicability to bio-fuels, their people would have free access to the lectures and seminars or even bringing in their own scientists to conduct research in complementary fields.

"I suggested they put up $100 million to buy into the Institute. There was no real basis for that number, we just made it up. Nobody at Exxon flinched. It was approved. The Exxon board of directors approved a general outline of the deal. It actually looked like it might happen. Roche backed out of it. The people in Switzerland got cold feet. They were concerned that Exxon might get close enough to see the level of profitability that was possible and decide to buy the whole company. They did not want to let them get a foot in the door, so they turned it down."

While that effort was dropped, it was clear the initial concept that had been born at a cocktail party less than five years earlier had successfully been transformed into a valuable entity; after only its few years in operation it was valued at greater than $100 million.

The entire atmosphere had been created to encourage and assist talented scientists to do great science, and take the lessons they learned at Nutley out into the scientific world. While the emphasis in Herb Weissbach's Department of Biochemistry continued to focus on protein synthesis and membrane biochemistry, Sid Pestka, Rakesh and Stephen Fairfield had begun conducting more extensive research into an odd protein nobody knew much about, known as Interferon. Interferon had been discovered only 15 years earlier by scientists working in England's National Institute for Medical Research. Because it was so difficult to obtain and isolate not much had been learned about it. Simply, it is a protein that is released by animal cells in response to the presence of a number of different pathogens, including viruses, bacteria, parasites and tumor cells. While of itself it does not kill any of those pathogens, it does fight them by signaling the immune system to go on the attack, and regulates the action of different genes involved in cell growth. Scientists were extremely intrigued by the potential disease-fighting capabilities of interferon, and some people began wondering if it might turn out to be a viable treatment for viral diseases and various forms of cancer.

Pestka's team had begun their efforts a year earlier, attempting to produce sufficient quantities of human interferon to permit them to begin conducting experiments by stimulating intact human organs. Unfortunately, as they reported, "So far we have been unable to obtain significant quantities of human interferon from human placenta by stimulation with either viruses or synthetic RNA molecules, such as poly(1)-poly(C). Although small amounts of interferon have been obtained...the major placental tissue did not produce any detectable interferon.

"We have begun to accumulate rabbit and mouse interferons in order to obtain sufficient material to purify and characterize it."

At RIMB the intellectual culture that was to make the Institute unique was slowly developing. The promises that had been made while recruiting scientists to join the staff were being fulfilled. As Nathan Brot would later describe it, "Because of the nature of the Institute, we could explore our academic and intellectual areas of interest far more than you might normally have been able to. Most important, at the Institute we had the opportunity to explore things outside our immediate areas, knowing that it was okay even if it led us down the wrong road.

"In addition to my primary projects, I was trying to develop a new assay. That was my strength; I had realized very early in my career that if you can develop an assay that is quicker and more sensitive than the guy next door, even though you may not be as smart chances are you're are going to beat him to the answer. A good assay gave you the advantage of time, which you can't buy. So I was always looking for a faster way of doing things. One of my projects was trying to understand the mechanism of how the ribosome was functioning in making proteins. The messenger RNA would bind to the ribosome and the amino acids would then be added on one amino acid at a time forming a new peptide bond in a sequence that was determined by the code in the messenger RNA. The ribosome was involved in the formation of each new peptide bond, thereby elongating the growing protein, but we did

189

not know what the ribosome was doing. We ultimate found that one of the proteins on the surface of the ribosome (L12) was involved in this process and I found that I could reversibly remove this protein from the ribosome and study it. During the course of these experiments I noticed that the L12 protein would rapidly lose its biological activity if it was subjected to mild oxidation.

"To solve this problem I listed all of the amino acids in proteins I knew to be modified by oxidation and to my surprise protein L12 lacked all of them. What was I missing? So very late one night I took out this old chemistry book and studied the chemistry of the amino acids. Finally I came across methionine, a sulfur amino acid, which I had worked with in the past, which can be reversibly oxidized to methionine sulfoxide and it was this oxidation that was inactivating L12. The oxidation of a methionine residue was also found to cause inactivation of other proteins, it is a general phenomenon. This, of course, could be lethal to the organism unless there was a method to reverse the oxidation. Thus we found a family of enzymes that is present in bacteria – and eukaryotes that could restore the activity of the oxidized methionine by reducing it back to methionine, and we were able to purify that enzyme. That turned out to be reasonably important because it potentially affects the activity of many proteins in which an important methionine is oxidized.

"It often isn't immediately apparent when a relatively mundane observation may have some significant importance. Science is always a progression of small steps. About the time I was working on this problem Roche was making recombinant interferon from E.coli. Herb Weissbach received a call from the company to discuss a problem they were having in producing pure human interferon in bacteria. Three peaks of protein were present in the purified interferon preparation. There was a concern that any modification of the protein might be antigenic, which would eliminate it as a drug. It turned out that the additional peaks were also interferon in which one or more of the methionine residues were oxidized. Our research in the Institute solved the problem and the purification procedure was modified to prevent the oxidation of the methionine resides."

In addition, years later it was shown that the oxidation of methionine residues in proteins may play an important role in the aging process and the enzyme we discovered, methionine sulfoxide reductase that reverses the effects of oxidation of methionine in proteins has received a great deal of attention in the aging process, as well as in other areas in which biochemical oxidation is an important reaction in the mechanism.

While the scientific reputation of the Institute was being firmly established, there also was an effort to make the place as collegial as possible. An important part of that unexpectedly turned out to be the Institute softball team. It was one of many activities that bonded the Institute together. The Institute fielded a team in the corporate league and, at least initially, nobody took it very seriously, a bunch of nerds playing sports. In fact, organizers ran a brief ad in the Institute newsletter, pleading, 'Wanted: Warm (hard) bodies to play (soft) ball. M or F.' Within the larger Roche Corporation the league was a big deal, there were two divisions, playoffs and the winning teams took great pride in their success. Everybody in the numerous divisions was aware of it and each different group would support its team. It was a very competitive league, and the Institute really couldn't compete. The Institute was only a small part of the giant pharmaceutical corporation so it had a somewhat limited group to draw players from – and scientists generally are not known for their athletic skills. As the Institute newspaper, the RIMB Informosome, admitted, "There were days this summer when cell biologists swarmed across the outfield at Brookdale Park like locusts over Egypt. One wondered how a ball hit there by a hopeful biochemistry batter could find the ground without first striking human flesh. Nevertheless, the balls did. Whether this phenomenon was determined by genetic composition or planetary motions cannot be explained – but time and again these deserving and enthusiastic fielders snatched defeat from the jaws of victory." So it didn't surprise anyone that the team went 0-12; when they won an occasional game those first few years it was considered a great upset. In fact, trying to recruit players for the team in the early years the Institute newsletter, challenged, "The second floor remains a big question mark, as it is unknown if anyone down there can throw a ball as far as second base... Active recruiting will begin any day now. Anybody who can run to first base faster than (scientist) Suresh Kerwar will be dragooned."

But in the late 1970s two postdocs with considerable baseball skills, Joe Tarnowski and John Kopchick, joined the lab. "I don't think we won a game the first season I was there," Tarnowski recalled. "We were something like 0-10, John and I didn't play very much the first season because we were the newcomers. They let us play three innings a game. But the next season they began letting us play – and we started winning. For years the dominant teams had always been the hourly employees working in the factories or in maintenance, they were the bigger more athletic people. There was a lot of luck involved. People used to joke that the Institute had begun screening post-doc applicants by their batting averages, but it was just a matter of getting the right people together in the right places – and for three years we were very good."

The team was nicknamed The Clones, although for the first few years most people referred to them as the clowns. They were the science geeks. When the team started winning the Institute rallied around them and the games became memorable social events. At times as many as 150 people from the Institute would show up at 5 o'clock. Sid Udenfriend, who previously had not been known as a baseball fan, showed up at almost every game; and he never missed a playoff game. Tarnowski remembers, "You would not know it was the same Sid Udenfriend. There was a pivotal moment in one playoff game. We were leading 2-1. I was playing third base. The other team had runners on first and second. The batter hit a long fly and our center fielder made a miraculous catch, and the runner on second tagged up to go to third. Our centerfielder threw the ball about six feet over my head. If the ball went over my head and into the stands the runners each would have been awarded two bases from their position – and both would have scored to win the game. I got down as low as I could and leaped as high as I could to snatch it. We made the play.

"Then I looked behind me – and Sid Udenfriend was right there backing me up. I have no idea what he intended to do, he couldn't play at all, but there he was with his arms outstretched."

The Clones won the league championship in 1981, and the success of the team really bound the Institute together. In addition to the softball team and all the clubs, there was a basketball league and a golf league and a volleyball league and there were picnics and card games and small and large dinners. The staff would regularly host what were known as 'international dinners' on a patio lounge, in which people would set up tables and offer the popular dishes that best represented their country of origin. The single largest annual event was a dinner to which every Institute employee was invited, from the dishwashers all the way up to department heads. The entire entrance floor became a huge banquet area and it was a beautifully catered event. The objective of all of these events was simply to bring people together in a comfortable setting and let whatever was going to happen, happen. Often what happened was science, as people traded concepts, information and theories.

The Institute – and the entire company -- maintained an active social schedule in an effort to bring people together. A Halloween variety show in 1974, for example, featured a comedy skit entitled 'Beat the Reaper,' in which game show contestants were subjected to a variety of colorful diseases. More than a year later 62 Roche employees, including many from the Institute, staged The Roche Players production of *Guys and Dolls* at a nearby high school for the benefit of a local charity. The Institute banded together to 'adopt' a young Bolivian child.

Certainly one of the most difficult relationships to work out and maintain was that between the Institute and the traditional pharmaceutical research laboratories. While the boundaries between academic and industrial labs had long been established and each carried with it well-known advantages and disadvantages, the Institute was attempting to create a different type of relationship. As someone once explained, it was close enough to the corporate labs that people on either side could look over the wall and see what was being done by those other scientists.

In the planning stage the importance of each area maintaining its independence had been emphasized, although no one really knew how that would play in reality. While openness and cooperation was being stressed inside the Institute, among the most human emotions are pride, ego and, at times jealousy. That is true even in the world of science, where a single discovery can be the basis of a long and successful career. Added to this was the fact that for the first time, with the creation of start-ups funded by venture capitalists, it was possible for a clever scientist to turn a discovery into a fortune. So there were unusual pressures coming from all directions to impact the relationship between Institute scientists and corporate scientists.

At the beginning the stated policy was strictly hands-off; that the Institute and the corporate side would do their own work under their own leader's direction. In fact, recognizing the potential for both benefits and conflicts, and determined to guard the independence of the Institute, Udenfriend proposed a set of guidelines to govern this relationship: "In some cases," he wrote, "our research programs have had a direct practical application which we agree should be fostered. However, we also acknowledge that this cannot be done at the expense of the basic philosophy that was followed in the development of the Institute: 'The Institute will be wholly devoted to long-term basic research designed to shed light on the fundamental life processes.'" To deal with this he proposed the guidelines that would be roughly followed; these included any collaborative project which would involve additional corporate support had to be presented to a member of the Executive Committee, it had to be initiated by an Institute scientist. It had to be for only a limited period of time and it could not represent "a major part of the senior investigator's program." In other words, on a severely limited basis, it was acceptable.

Most often the relationship between the two entities was unusual and at times controversial. Herb Weissbach remembers having a conversation with Roy Vagelos, whom he had known when both were working at NIH while later Vagelos had become President of Merck. Vagelos was wary of the structure of the Institute. "He wanted to know how the Institute was set up, what its philosophy was and how we operated," Weissbach said. "I explained it to him and he shook his head and said he wouldn't do it that way. He would want the scientists doing basic research working side-by-side with scientists inside the company.

"It was exactly the opposite of our philosophy – and the fact that we were able to do it was one of the reasons we feel the Institute was so successful for so long. There is no right or wrong. Our goals were different and we had such a different culture than the corporate side that it couldn't have worked for us. For the Institute to work we had to have our critical mass in one place with easy access to everyone else, to promote interaction."

In fact, Irwin Learner, who served as Roche President and CEO for more than a dozen years, had little knowledge of what was going on inside that big modern building. "It wasn't important that I knew," he said. "I had no power to influence or control what they were doing anyway. I had tremendous faith and confidence in John Burns, so as long as he was satisfied I was satisfied.

"My most important function was to make sure that there was no open warfare between the two research facilities. But in fact that was never something I had to be concerned about. Very quickly they developed a good collegial scientific respect on both sides. The Institute management did a remarkable job never allowing anyone to believe that the Institute scientists looked down on the scientists on the product development side. At any of the open lectures or presentations everybody was treated with equal respect; there was never a feeling that the product people were any less important than these brilliant scientists; it was much more of collaborative feeling, that everyone was operating on level ground, equals among equals. "I think after a while the concept took hold that everyone there was part of a great company."

Lerner contributed to building that relationship by making himself highly visible to Institute employees and continually offering encouragement even, as he admits, if he didn't have the slightest idea what they were doing. "Roche even at that time was a very large company. In addition to the pharmacology sector, our production facilities and the Institute we also were the world's largest producers of vitamins. I spent a considerable amount of time visiting every corner of our operations. I'd walk around our campus for five or six hours a day. When I went into the Institute I'd talk to a lot of our young scientists. I had no idea what they were doing, but honestly, I don't think many of them knew where their research was going to lead them. But I wanted them to know that I was there and emphasize the fact that they were part of the Roche family. While they didn't impact the commercial side, the Institute was an intricate part of what made Roche unique as an organization; Roche was a company that had the courage and innovative spirit to create such an entity in what had been a totally commercial enterprise.

"When people asked about the cost-benefit I told them it couldn't be calculated in dollars and cents. The Institute gave us tremendous credibility in the science world and the best scientists were not at all reluctant to come to Roche to pursue their own careers. The Institute itself brought more and more bright young scientists from places like the NIH and enhanced the entire company.

There were many attempts to make that wall porous. People inside the Institute did get to know their counterparts on the corporate side, especially during the first few years when Institute scientists shared corporate space, and there were areas of collaboration. When it was possible the Institute would use the corporate technology, and whenever it was practical share scientific information with Roche scientists. According to Science magazine, "(R)esearchers at the Institute are afforded access to many of the facilities of Hoffmann-La Roche's massive applied research enterprise, including x-ray crystallography, mass spectrometry, nuclear magnetic resonance, large scale preparatory facilities – and a farm for large animals for immunology. Institute scientists need only submit a memo to the company's organic chemistry division for synthesis or analysis of unique compounds."

"They had established capabilities that we didn't have and they shared them willingly," Herb Weissbach says in a complimentary fashion. "To conduct our research we needed huge amounts of bacterial extracts. The genetic code is the same in all living things. We worked with bacteria and John Burns, based on my request when I came to Roche, established a biopolymer laboratory that eventually was housed on the 5th floor of the Institute building. They could easily produce pounds of the E. coli bacteria for us so we could purify all of the protein factors required to convert the information in a gene to a protein in a test tube." Later the biopolymer lab was important for the large scale production of interferon when Roche moved into the field of recombinant DNA.

What made the relationship complex was that the goals, the expectations, working conditions and the objectives of the two entities were totally different. The decisions as to what scientists inside the corporation would work on were made by management and they most often came with a timetable. They were told what they were expected to accomplish within a period of time. They had to continue moving forward to meet those set schedules, while scientists inside the Institute made their own decisions and set their own timetables.

The right to publish your discoveries were another significant difference. The Institute charter gave scientists the right to publish their discoveries. That was essential because they wouldn't have joined the Institute without that right and, in fact, few of their discoveries had any current commercial value. That was much more difficult inside the corporation where they were developing specific products. Everything those scientists did had to be protected to insure corporate ownership.

It took some time to establish a truly cordial working relationship between the two entities. At the beginning people working inside the corporation looked at this new building up on the hill and had not the slightest idea what was being done there. Clearly some people wondered why Roche was forming this new and separate entity – and perhaps some people felt threatened by it. At the beginning the Institute allowed Roche to take an important step into the future, without interfering with ongoing and very profitable operations. "In 1970 to most scientists molecular biology was a bit of a mystery," Arthur Weissbach remembers. "If you were trained as a biochemist or a chemist you did not have any exposure to molecular biology beyond knowing what DNA was. And the emphasis that DNA people or RNA people put on the informational content of the molecules was alien to them. But they wanted to learn, and we eventually made some good friends over there.

"There certainly were people in the company who didn't like the Institute. People inside the Institute had the feeling that they were individual entrepreneurs; there was a sense that everyone was running our own business. But in the company they were working on assigned projects. How much envy there was because of our freedom I couldn't quantify. I think that was always a sticking point. I remember two separate conversations with corporate scientists who were concerned about it. However, the reason I had these conversations was that we were collaborating with them, which sort of diminished the importance of their complaints."

But remarkably, the artificial wall disappeared rather quickly. Eventually, as Udenfriend reported in 1981, "Our association with Hoffmann-La Roche brings us more than direct financial support. Research and development at Hoffmann-La Roche consists of 1300 scientists, including some 300 at the doctoral level. The interaction of our members with those scientists has been most important in fostering our scientific excellence. The interaction ranges from collaboration on research problems...to important technical assistance, such as assays by analytical laboratories, synthesis by organic chemists, large scale isolations by the biopolymer laboratories and preparations by the fermentation group. Most important, our presence on the Nutley 'campus' permits Institute scientists and scientists from Hoffmann-La Roche to meet on a day-to-day basis, thus providing a convenient framework for the interchange of ideas."

In addition, Udenfriend continued, "Important methodology and instrumentation developed in the Institute has been applied to research and development projects of the company."

That collaboration didn't happen especially quickly or even easily. But within a few years that interaction between the scientists to which Udenfriend referred had taken root. According to John Burns, by 1971 "Collaborative programs are under way on new approaches to cancer research and to mechanisms involved in narcotic addition." One of the first collaborations from which the company actually saw a benefit was the purification of proteins and peptides. In 1974, for example, Roche was looking for new methods of identifying and purifying compounds that existed in nature, and some of the basic research done at the Institute had laid some of the groundwork. Udenfriend's research, in particular, led to novel ways of purifying trace amounts of proteins and peptides. So scientists began collaborating to isolate, purify and sequence biological peptides and proteins. Each facility did that aspect of the project they did best but as a result Roche was in the forefront of protein and peptide purification. The corporation eventually was able to obtain patents on several products, including thymic peptides – a lot of work which ultimately matured into the interferon project.

Herb Weissbach noted that there seemed to be an unwritten agreement that the Institute was not going to steal scientists from the Roche research group, but movement in the other direction occurred frequently. Art Weissbach pointed out "As my Department matured there were some people working in my Department who wanted to go over to Roche and establish a new kingdom, a new research group within the company."

One person who did attempt to transition from the Institute to the pharmaceutical building was Anna Marie Skalka. Skalka had been one of the original scientists recruited to work in Arthur Weissbach's Department. After spending "five happy years at Cold Spring Harbor as a fairly independent postdoctoral associate supported by the Carnegie Institution of Washington..." she was hoping for a permanent position. But when James Watson shifted the emphasis of his lab to the study of eukaryotic cancer viruses, she began looking for another position. "I received some nibbles and I interviewed for faculty positions in a few university departments. However, I was shocked to hear from a female member in one of the departments that while I gave a terrific talk I should not expect an offer as the department already had 'their token female.'" Instead, she accepted Art Weissbach's offer to become a founding member of the Institute.

Initially the focus of her work was the bacteriophage – a virus that infects bacteria rather than human cells -- Lambda, which Berg had used in his gene splicing experiments. As she had learned from her mentor at James Watson's Cold Spring Harbor Laboratory, Dr. Alfred Hershey, "Understanding a class of phenomenon is understanding a class of phenomenon – or it's nothing.

"We were studying a simple system. Lambda is a virus that preys on bacteria. We were very interested in understanding what its structure was and how that structure related to its function. It was giving insight into the way other systems function. It was simply a building block, but it enabled us to understand many of the other building blocks. It turned out to be a very important factor in cloning genes of higher organisms. It put me in a very good position to enter the whole field of recombinant DNA. I actually taught one of Sidney Pestka's postdocs how to clone – and that enabled Pestka to clone the interferon gene, which was the origin of Roche's first product from recombinant DNA."

Eventually Skalka decided she wanted to shift her focus from this virus to viruses known to cause cancer in animals. The reason, she explained, was that "The field was becoming more narrowly focused. It was becoming more and more difficult to attract good postdoctoral fellows. I began looking for another area of research. I became fascinated when it was reported that a covalently closed, double-stranded DNA circle was produced during retroviral replication... I decided to close down my phage work and become a retrovirologist.

"This rapid switch in fields would have been difficult, if not impossible, if I were at a university and depended on grants for my research." Skalka continued. "But support and encouragement for me in this change was a great vote of confidence and a special advantage of our unique institution." With Art Weissbach's approval she took a three month leave of sabbatical to study retroviruses and returned a retrovirologist. Because of this expertise at one point John Burns recruited her to cross the campus and go into the company to set up a group that would be studying cancer genes. "It was not something I had wanted to do," she remembers. "I felt very close to my colleagues inside the Institute. But they prevailed upon me to make the transition so I did. The attraction to me was to build a group that would do important science on cancer genes. They made me an Assistant Vice-President and we created the Department of Molecular Oncology. I don't think my compensation was very different. But John Burns painted a very nice picture about me starting a group that would be doing important research into cancer genes. So finally I agreed to make the transition. I recruited a number of outstanding retrovirologists and we went to work.

"It was like walking into a totally new world. The psychology was totally different. The goals of the individuals were totally different. There also was a lot of jostling for positions, there were political decisions to be made, so it was no longer about the science. The goal was to make a drug, not gain an understanding of fundamental life processes.

"Once I got into the company they assigned me additional duties. I was placed in charge of their biotechnology group and had very different responsibilities. Then John Burns retired and what I was doing was not so important to the people who replaced him. After two years our Department of Molecular Oncology was incorporated back into the Institute and I was able to go home."

After Nikolaos Robakis had completed his postdoctoral fellowship at the Institute he was offered and accepted a position inside the company. "It was a completely different culture and I wasn't comfortable there," he said. "The company was doing a lot of good work, but it was much too much of an engineering job for me. Roche was the leading producer of vitamin C at that time, and they were looking for a way to genetically engineer bacteria so that we could metabolize or produce vitamin C simply. We had to focus all of our efforts on that. I didn't like those restrictions, so eventually I left."

But there were several instances of people successfully making the transition from pure science to product development. When the company formed its Department of Molecular Genetics, for example, Dr. Pat Gage, who had been working in the Institutes Cell Biology department, was appointed a Vice President of Research in the Company and took a sizeable number of his staff down the hill with him.

But few people saw both sides of the company as well as Joseph and Deb Tarnowski. When Tarnowski finished his graduate studies at the University of Tennessee he accepted an offer from the Institute to continue his post-doctoral research because "Like a lot of people I was torn between academia and industry. I had watched my professor struggle with grant writing and publishing or perishing and boy, it was a lot of rejection. But going into industry was still considered selling out. The Roche Institute seemed to be some kind of unique amalgamation of the two.

"They had specialties in cell biology, protein chemistry, molecular biology and they were seeking post-docs to come and work with investigators." After Tarnowski was offered a position his wife applied for a job as a lab technician in both the Institute and the corporate lab – and received offers from both entities. "She accepted the job in the company, which offered benefits I didn't have as a post-doc."

While Deb Tarnowski was working on a project to isolate interferon, Joe Tarnowski was working with the DNA in spinach. "The goal of the program I was in was to make an original discovery," he says. "I was isolating chloroplasts, the green material in spinach leaves, to look for transcription factors – the molecules that regulate DNA by turning genes on and off. My research materials came from a market in Paterson, New Jersey once a week. My expense account included $5 for spinach. I was looking especially for anything related to light activation. It seems like it has to be there because if you grow a green plant in the dark it doesn't turn green, it remains sort of yellow and white. We were trying to figure out the activation factors and to do that I had to fractionate my materials."

The fact that Tarnowski was working in the Institute and his wife was working in the corporate labs gave them a unique basis for comparison. Rather than being competitive, Joe Tarnowski found, "It was very collegial – but there was some jealousy. Inside the company they referred to Building 102, the Institute building, as the country club, because they thought we were accountable for nothing and had an essentially unlimited budget while everybody else had to deliver something and live within strict financial parameters."

Tarnowski stayed at the Institute for about 14 months. The whole concept of the post-doc program was that a postdoc would stay only for a limited period of time, then go work somewhere else as a scientist. For Tarnowski, that somewhere else turned out to be down the hill. His wife's group was developing Interferon as a therapeutic and their initial data were looking very good. They wanted to expand that research and were hiring scientists who had a molecular biology background, in particular protein chemistry.

"When I told Herb Weissbach I had applied for the job he supported me completely. I was hired and made the transition. Ironically, we still were working at the Institute because all of the process development and clinical trial manufacturing was being done on the 5th floor of the Institute. So there was a lot of collaboration. What I found was that most scientists care about doing good science, and they respected the people who did it. So rather than any real jealousy or competition, people just did their jobs; when they needed assistance it was available and most people I knew took advantage of it."

The Institute was also valuable to the company in another way. As its reputation grew, it helped change the overall image of the Roche for young scientists who initially looked down upon corporate laboratories. Roche President Irwin Lerner said, "The Institute was immeasurably valuable in helping us attract great scientists. The credibility of Roche and its devotion to science improved and increased dramatically all around the world. It was clear that this was a different company compared to the others in the industry, the Lilly's and the Merck's and that image made a significant difference in the overall success of the company."

In retrospect, rather than creating resentment, most of the people working at Roche came to respect the Institute and were proud to be part of the company that created it. The success of the Roche Institute arguably had an impact on the entire industry. In his 1981 Director's report Sidney Udenfriend wrote. "This year I have chosen a subject that has generated great interest as well as concern, namely the new alliance between industry and academia. Within the last two years a number of companies have concluded arrangements to support ongoing research in university laboratories, often at levels in the multi-million dollar range. Obviously industry seeks something in academia that is lacking in their own research laboratories. I believe the relationship between the Roche Institute of Molecular Biology and Hoffmann-La Roche may provide the model that they hope to emulate."

Chapter Five

The Institute Becomes a World Class Scientific Center

By nature, scientists tend to be quirky. Which certainly is among the many reasons so many scientists spend their lives exploring the mysteries of nature. The men and women who came to work at the Roche Institute were typically unique representatives of scientists provided the ultimate tools of researchers – time, complete intellectual freedom and money. There was no such thing as a typical work week or even a work-day. It was common for people to be finishing their day while others were arriving to begin work.

Almost from its inception the Institute began attracting truly committed scientists. Reflecting on that, Arthur Weissbach said persistence, commitment, curiosity and passion were the common traits he had seen in the scientists he most respected. "It seemed to me that the best scientists lived with their work and their research 24 hours a day. They are always thinking about it;

they are engrossed in their project. That can be deadly to their family because it can cause them to ignore their responsibilities or not handle them right.

"But that tends to be the exception. It's possible to be a make-a-living scientist. In fact, there are many scientists who have no desire to make a total commitment to science. They are very satisfied working for a strong leader who tells them what to do. That's fine with them, they are not entrepreneurs. For them science is a 9-to-5 job and they don't want their work to interfere with the rest of their lives.

"We had far fewer of those people at the Institute. That certainly was one of the things we looked for when building our staff. We tried to hire very strong leaders, people with passion, people whose youthful enthusiasm had matured into a passion."

Added to the already varied brew of personalities was an extraordinary number of nationalities as people from around the world joined the staff. Because post-docs were permitted to stay at the Institute generally only three years, there was a continual influx of people. The operating concept was that these people would be trained by a group of outstanding scientists, then take their acquired skills to laboratories around the world. While no one ever totaled the number of countries represented in the lab, or the primary languages, these included scientists, postdocs and researchers from England, France, Germany, Japan, China, Greece, Scotland, Sweden, Yugoslavia, Israel, Poland, Chile and India, Pakistan, Greece as well as many South American countries. It included people from just about every continent – except, naturally, Antarctica and at times there were scientists and researchers from more than 14 different countries on the staff.

Among the first immigrants to join the Institute were two scientists who escaped from Communist Czechoslovakia. At a meeting of the International Congress of Biochemistry in Prague in the late 1960s, Art Weissbach quietly had been approached about helping scientists get out of that country. He managed to arrange an appointment for one member of the Czech Academy of Sciences at Columbia University but for others the route to freedom – and Nutley – was far more perilous. Biophysicist Paul Bartl and his wife had driven to the west with permission – but their 5-year old daughter was hidden in the trunk of the car. They were forced to leave Bartl's aging father behind. Electron microscopist Milos Boublik escaped in 1971 and it was his decade long collaboration with scientist Frank Joachim that led, as Joachim wrote, "to the first three-dimensional reconstructions of ribosomal particles."

But quirky was the norm, remembered Tim Coleman. Typical of the unique researcher who worked there was a postdoc,

Bob Shoeman. "He was absolutely brilliant, but different. He spoke with the exactness of a computer. He had a photographic memory and could read an entire scientific paper in three to seven minutes. But what I remember most was his filing system. After reading a paper he would drop it in a box. Months later I might ask him for a reference he'd mentioned to me about an experiment and he would pick up a box, turn it upside down on his desk, cut the pile like a deck of cards and hand me the paper. He could never explain how he did it, but he did it.

"To me, he was the essential curious scientist. He was a hunter who was not satisfied with commercial ammunition so he made his own bullets. After he'd bought his first car he wanted to understand how it worked, so took it apart, down to the nuts and bolts, then put it back together. He did these things because he could. And he was the same way in a laboratory.

Of course, not everyone fit well into the Institute. Anna Skalka remembers that the scent of marijuana seemed to surround the desk of her first post-graduate assistant, and that he left after only about three months – to join a commune.

Any doubt that the Institute had become a desirable place to work was settled forever in 1974. Udenfriend and Herb Weissbach decided that the final prestigious piece they needed to add to the staff was a Nobel Prize winner. They knew of two Nobel recipients then living in New York. The first, Fritz Lipmann, had been awarded his prize in 1953 for "his discovery of coenzyme A and its importance for intermediary metabolism," and the second was a Spanish born scientist, Severo Ochoa, who had shared the 1959 award in biochemistry with Arthur Kornberg "for their discovery of the mechanisms in the biological synthesis of ribonucleic acid and deoxyribonucleic acid". Ochoa was then Chairman of the Biochemistry Department at NYU.

Udenfriend offered Lipmann an attractive compensation package, but eventually Lipmann asked for too much: He wanted to be chauffeured back and forth to his apartment in New York to the Institute in Nutley. But as Roche Vice-President John Burns pointed out, no one had that benefit and they couldn't add that to his package. Ochoa was always a more likely choice; He and Sid Udenfriend had met almost 30 years earlier, when Udenfriend was a post-graduate student in biochemistry at NYU Medical School. Also his research on protein synthesis was in the same field in which Herb Weissbach was working.

And while it was Lipmann who wanted a driver, it actually was Ochoa who could have used one. "He loved to drive in his Cadillac," Herb Weissbach remembers, who would on occasion

218

drive with him, "but he was not the best driver. He reminded me of that character, Mr. Magoo."

Ochoa brought his reputation with him. As the Institute's newspaper proclaimed, "His presence in the Institute's Biochemistry Department... will establish this as one of the major centers for research in protein biosynthesis." In the 1950's he had successfully synthetized RNA using a bacterial enzyme called polynucleotide phosphorylase, which received much attention in the early 1960s since that enzyme could be used to make the synthetic RNA polymers that helped crack the genetic code. His lab at the time was competing with the Nirenberg lab to decipher the code. His decision to join the Institute only seven years after its inception gave it that final stamp of scientific approval. Udenfriend and Herb Weissbach's main concern was whether he would be comfortable in this new, young Institute that was part of a pharmaceutical company. His presence, Weissbach remembered, was the final stamp of approval that this was going to become a respected place to do research. "He was 70 years old at that time and we didn't expect him to actually do any research. But he came in with his research group and spent the next decade doing fine work. He turned out to be a wonderful colleague, warm, easily approachable and clearly an outstanding scientist. He regularly attended seminars and lectures and participated in the journal clubs, becoming one of the guys."

As Richard Snyder recalls, "Ochoa was a most helpful individual and an absolute gentleman. He occasionally would come down to see me and he'd say, 'Dick, I made a personal phone call to South America. I want to pay for it.' Few people would have gone to that length for something that easily could have been related to science. Really, he was an amazing man."

Arthur Kornberg remembered that when he was working with Ochoa, a lab assistant had spent several weeks purifying sufficient malic enzyme from pigeon livers to be able to conduct an experiment. As he and Ochoa were pouring the final fraction into a measuring cylinder he slipped and knocked over several bottles, which tipped the cylinder – causing it to empty its contents on the floor. The experiment was ruined.

Kornberg was despondent, he recalled. Ochoa appeared to take the setback in stride, assuring him everything could be replaced. Later that day the concerned Ochoa called Kornberg at home – twice – because he was concerned about his colleague. Fortunately, Kornberg recalled that he had saved and stored an earlier fraction; and when he assayed it he discovered it contained most of the enzyme activity – saving the experiment.

At the Institute Ochoa continued the research he had been conducting at NYU: Protein Biosynthesis, Molecular Mechanism and Regulation, which he described as, "using embryos of the brine shrimp, Artemia salina, undeveloped or after partial development, to study molecular mechanisms of polypeptide chain initiation in eukaryotes as regards (a) the number, nature and mode of action of the initiation factors, and (b) translational control mechanisms."

It simply was another achingly small step in the effort to understand how nature works. In every lab in the building similarly small steps were being taken to unlock the secrets of the cell and life itself. It had taken only a very few years to progress from cocktail party chatter to a world-class laboratory in which cutting-edge research was being conducted on a daily basis.

While eventually the Institute would be able to boast that seven National Science Academy members and a Noble laureate were all working on its four floors – there were so many that Amiya Banerjee would joke, "If you'd like to meet a National Academy member just get on the elevator and get off at any floor." But as Arthur Weissbach remembers there was at least one great scientist who got away. When he was creating his Department of Microbiology, which very quickly was renamed the Department of Cell Biology, one of the first senior staff people he hired was

Gebhard Koch, whom he'd met at NIH. Koch was working on poliovirus. Arthur Weissbach's lab was focused primarily on the question of how does a virus get into a cell and how does virus DNA become virus RNA? The mechanism by which RNA worked was a fascinating mystery, which included the fact that some RNA viruses were retroviruses, meaning their genetic information contained in RNA could be rewritten into DNA, and then incorporated into the DNA of the cell. It was exactly the opposite of most of nature, and Arthur Weissbach's lab wanted to better understand the process. The thought was that it made sense to bring in researchers with a variety of expertise in cell biology, rather than maintaining a focus on virology.

"When Gebhard Koch came in," Weissbach said, "he already was very famous in his field so we gave him one of the largest allotments of laboratory space, as much as anyone else in the entire Institute.

"He did excellent work and a few years later wanted to bring in a scientist named Mike Bishop, who had worked for him at the University of Hanover. At that time the entire field was expanding greatly and I was concerned about diversity – meaning I wanted to bring in a developmental biologist or an embryologist. I had my heart set on it. Mike Bishop had been a post-doc with Gebhard and I assumed he was a virologist and I had already decided I wasn't going to hire another virologist. So I turned him down.

"Years later J. Michael Bishop shared the 1989 Nobel Prize in Physiology or Medicine with Harold Varmus for their work on host genes involved with cancer. It was a revolution, because it allowed other researchers to investigate how cancer genes within our own DNA were controlled, and what turned them on or off.

"Even decades after that decision I was still kicking myself for it. The important lesson I learned from that is that I should have gone with what was working well, rather than trying to jump into an unknown field."

While producing marketable products was never a goal of the Institute, its potential was shown very early in its existence. Syd Spector's immunology project was the first clear example of what Roche was hoping for when the Institute was established. Although not stated in the charter, in a sense it was the first proof of the charter concept, namely, that if you provide good scientists with the resources they need to do their science, new products will be developed.

Would a scientist actually have that promised complete freedom to chase his own rainbows? "My role as Director of the Institute," Udenfriend once wrote, "is to maintain a high level of scientific quality. I do not direct the research of its independent scientists." There was never any question that Udenfriend fully supported Spectors' research program.

Spector was a firm believer that "New methodology moves science," and so he adapted a relatively new technique at that time, radioimmunoassay; which allowed researchers to quantify minute amounts of biological substances by using radioactive material, to his research. Spector developed antibodies to several drugs so he could study their metabolism. These included antibodies to a wide range of drugs, including barbiturates, morphine, reserpine, and several others, in addition to neurotransmitters, serotonin and acetylcholine. The actions of the antibodies that he produced intrigued him; he noticed that "they have characteristics that are very much like receptors, they had specificity and high affinity, which made the assays very sensitive. I decided to go on a fishing expedition and wondered whether one could use antibodies as surrogate receptors, something like a fishing hook?

"I found that antibodies had a greater degree of specificity than the cell receptor... It's incredibly sensitive. You can get down to measure nanomole levels. The other advantage is that you can assay a much greater number of specimens with radio-immunoassays than with other methods." Spector had developed this assay to assist in his own research and had not considered its potential commercial applications.

Following the protocol that was being established, Udenfriend sent Spector's soon-to-be published paper concerning making antibodies to morphine to corporate attorney George Gould to determine if the company had any interest in pursuing a patent. It was meant to be purely perfunctory; no one inside the Institute saw this as a product. "I just had the funny feeling it should be evaluated by somebody," Gould remembers. "I sent it to a scientist in the Roche research department. He got all excited by the opportunity to manufacture a diagnostic assay against a drug of abuse. We decided to go ahead and do a quick patent filing before the publication came out. That became the basis for the first commercial product to come out of the Institute, called Abuscreeen." Abuscreen was a relatively simple test that was used to screen body fluids for the presence of a variety of different drugs with a sensitivity that had never before been possible, and was commonly used in industry to test employees for drug use. It eventually became Roche's Diagnostic division most successful product.

Spector continued his work in this area and soon made another important and unexpected discovery: The fact that the human brain produces morphine. His experiment, as he explained, followed Udenfriend's rule, make it simple: He put radioactive morphine into an extract of brain tissue and searched for any substance that competed with the morphine for the available receptor sites on the antibody. To the surprise of Spector and his team, they discovered a mysterious substance in cells capable of bonding with the same receptor. They began by finding out what it wasn't, demonstrating it wasn't a peptide. In fact, the only way to prevent this competition for the receptor by the natural substance was to oxidize it with iodine – which is exactly what would occur with morphine. "It looked like morphine," he said, "but we hadn't proved it yet. We had to go through a series of purifications, and sure enough, it came out as morphine."

They then proved that it was naturally produced morphine, rather than something coming from an outside source. Spector proved that the body produced it, but not how it was used. His theory, he said, was that "When we are sick, the physician in the body calls on these drugs, the pharmacist in the body prescribes these drugs and the physician in the body gives the right dose. The good Lord has provided us with all these key components." The actual purpose and function of this morphine has remained a mystery, although there has been some speculation it plays a role in the surprising fact that placebos often have an important physical impact on the body. But more than four decades after this initial discovery the journal Neuropsychopharmacology wrote "The potential of (Spector's) studies is just beginning to unravel."

It had taken the Institute only a few years to prove both its value and its potential, both to Hoffmann-La Roche and the scientific community. But it had been launched at an extraordinary time in scientific history. A new world had been discovered, a world in which it was suddenly possible to manipulate DNA and alter existing life -- and with that came the possibility that at some point in the future scientists could create new life forms. This new capability brought with it both moral and ethical questions about the way this technology might be used. How far was too far? As this was precisely the type of research many of the scientists at the Institute were engaged in they had a real stake in the outcome. "When it became clear that there was now the technology to actually clone any gene from any organism, human, animal, or even a virus from any organism, people got scared," Anna Skalka remembers. "They thought, oh my God, what are we doing? We're putting genes from pig cells into human cells. They got frightened that scientists might do something crazy like pick up a toxin gene and put it in a human virus."

In fact, the 1968 book, The Biological Time Bomb by British author Gordon Taylor drew an extraordinary picture, "You may marry a semi-artificial man or woman...choose your children's

sex...tune out pain...change your memories...and live to be 150..."
but then he added ominously, "if the scientific revolution doesn't
destroy us first."

Science had advanced to the point at which it actually was
possible that someday a Frankenstein monster could be created.
As Arthur Weissbach added, "A lot of people, especially the
religious community, got very upset. They had heard that
recombinant DNA technology was an affront to God. God didn't
make it, so it had to be bad stuff. So in response to the concerns of
the scientific community, as well as the public, a conference was
held in February 1975 in Asilomar, California. This Asilomar
Conference, as it is known, attempted to set up some guidelines.
All the players in the recombinant DNA community attended this
conference." Udenfriend sent Anna Skalka to represent the
Institute.

In some ways, the attendees were designing the future,
discussing what was possible to do with this technology, and if
there was a possibility of harming the environment, what
reasonable precautions the scientific community should take.
There really had never been anything quite like this before in the
biomedical field. Among the few comparable scientific advances
was the creation of the atomic bomb, although rather than civilian

nuclear scientists having control of that technology, because it was the result of a military effort, its use became a military decision.

The Asilomar conference is considered one of those meetings that had a profound impact on the scientific community and the government's response to this new technology. It had been organized by Paul Berg and Maxine Singer, who were working together to make some of the breakthroughs in recombinant DNA research, both of whom Arthur Weissbach and Herb Weissbach had known. As Berg said afterwards, "The people who sounded the alarm about this new line of experimentation were... scientists. They had called for a worldwide moratorium on the work, followed by an international conference of experts at which the nature and magnitude of the risks could be assessed... The conference marked the beginning of an exceptional era for science and for the public discussion of science policy...

Berg also noted that "Most researchers, like me, acknowledged that the new technology opened extraordinary avenues for genetics and could ultimately lead to exceptional opportunities in medicine, agriculture and industry. But we conceded that unfettered pursuit of these goals might have unforeseen and damaging consequences on human health and Earth's ecosystems."

About 140 people attended the conference. It was, as Skalka recalls, a unique gathering of many of the leading scientists in the world. Their goal was to plot a path that allowed the research that needed to be done to take place, but to do so within guidelines that made it safe. There was, as would be expected, a lot of disagreement and debate among these scientists as to how much risk actually existed in this technology. The assessment ranged from no risk at all to disastrous. The fear that pervaded the conference was that if the scientific community did not act voluntarily, government might step in. That, of course, was what happened decades later with stem cell research.

One of the leading critics at the Asilomar Conference was Noble Prize recipient, biologist Joshua Lederberg, who circulated a paper arguing that rather than being restricted, this new technology would make amazing discoveries possible. The ability to recombine genetic material offered "an early chance for a technology of untold importance for diagnostic and therapeutic medicine: the ready production of an unlimited variety of human proteins. Analogous applications may be foreseen in the fermentation process for cheaply manufacturing essential nutrients, and in the improvement of microbes for the production of antibiotics and of special industrial chemicals."

On the last day of the conference an agreement on guidelines was reached. For example, researchers were limited to using certain strains of the E. coli –known as 'the bicentennial bug' because these rules were promulgated in 1976 -- that were difficult to grow, insuring they would not spread easily. Risk estimates were assigned to various experiments which dictated how they would be conducted. Experiments believed to have little or no risk were permitted to be done on an open bench and as the potential was judged increasingly risky the safety procedures were made more stringent. It was agreed that there would a moratorium on research – it lasted 16 months – during which the NIH would create these guidelines. Dealing with these restrictions caused a lot of unnecessary problems, Skalka remembers. "There was a recombinant DNA committee set up in the NIH to rule. We would have to go to them and explain what you wanted to do and they would either allow you to do that experiment or turn you down. My goal was to clone retroviral genes. These rules put me completely out of it. So I had to scratch my head and decide what I was going to do. I wanted to clone avian sarcoma viruses but I couldn't do that, so I decided I would practice by cloning chicken genes.

"I decided to clone some ribosomal genes from chickens, but chicken genes were judged very dangerous, so in order to do this I had to go to a special field 4 facility in Building 41 at NIH. I had to strip down completely, shower, put on a jumpsuit and go into this facility and work in glove boxes, cloning chicken DNA!

"A post-doc and I went down and spent three days going through all of this. I made all the fragments and cloned all the genes into lambda, put them in a paint can and sealed it. I had to dip the paint can out of a bath of Clorox and I took this back to Nutley and had to store it in a cold room until the moratorium was lifted. It was three years before I could open that paint can and get to my chicken DNA."

Many scientists believed these rules were completely unnecessary, and after a few years they were greatly relaxed as further experiments revealed, as Berg wrote, "the fear among scientists that artificially moving DNA among species would have profound effects on natural processes has substantially disappeared with the discovery that such changes occur in nature."

In the decades since the Asilomar conference there hasn't been a single incident of public health being endangered by experiments with recombinant DNA. Whether that is due to the voluntary steps taken by scientists or would have been the inevitable result even without those precautions obviously never will be known. Paul Berg and Maxine Singer both believed the conference accomplished several important objectives, among them making the genetic research being done in laboratories transparent, and perhaps most important making it completely unnecessary for the government to regulate this potentially world-changing experimentation.

The attention the conference brought to the new world of genetic engineering attracted the interest of the new industry called biotech, then in its embryonic state. Through the years great fortunes had been made by people leaping in early to take full advantage of newly discovered technologies, and in several instances supplying the necessary financial support to aid in that commercial development.

The birth of biotechnology as a multi-billion dollar industry can be traced to the early 1970s when Cetus Corporation was established near Berkeley, California, primarily to find micro-organisms that would allow it to greatly enhance the process of producing antibiotics and vaccine components as well as chemical feed stocks. But in 1976, just as the Asilomar moratorium was expiring 28-year old venture capitalist Robert Swanson asked Herbert Boyer to meet with him for 10 minutes. That meeting instead lasted several hours and changed the scientific world in fundamental ways. Genentech was founded. As Frederick Dorey reported in the Wall Street Journal in 1999, "It is not too much to say that this meeting produced a new structure for science-driven business and fundamentally changed the relationship between business and America's research universities...

"Swanson...wondered whether new organisms could be created to help discover and produce pharmaceutical proteins. The answer seemed to be yes... Swanson became CEO of Genentech and set out to create an organization that every scientist would want to join. The quality of their science was paramount; they could publish as much and as often, as warranted. Genentech developed its own products and retained the commercial rights. Genentech protected those rights through patents, and reported their findings. Through stock options scientists would share in the success of their work." Four years after its founding Genentech went public and its IPO was at that time the most successful initial offering IPO in Wall Street history.

David Goeddel, the first scientist Boyer hired at the new company explained their corporate philosophy, "No matter what we were working on we knew that if we didn't get there first, Genentech might not survive. Each project was the same; we've got to be first."

As part of their compensation package, scientists at Genentech were offered stock in the company. That marked a fundamental shift in scientific research: For the first time in history the scientists who were doing the research could become wealthy by sharing in the corporate profits. The creation of Genentech gave scientists a third potential career path to follow: they could pursue pure research in a university lab, work for hire in a corporate lab, or join one of the new scientific start-ups that had begun springing up to take advantage of this new technology of genetic engineering and possibly earn a fortune.

The financial success of Genentech caused investors to rush into the world of biotech. For many young scientists the chance to make a lot of money while conducting cutting-edge research was an enticing possibility, and in at least some instances forced them to examine their own objectives. "Suddenly there was money to be made working for a start-up," Herb Weissbach points out. "They were looking mostly for experienced scientists who offered them the best chance of creating marketable products, so the lure was there for people on our staff to leave. I was dismayed that Genentech later hired a scientist from our Institute and one from the Roche biopolymer laboratory. Another example, unrelated to Genentech, was Craig Rosen, a brilliant young scientist who I appointed Head of the Biochemistry Department, but eventually left the Institute to help found Human Genome Sciences. Through the years though, we were fortunate that our scientists remained happily at the Institute and the loss of our staff to other companies never became a significant problem for us."

An issue that might have become a problem, but rarely did, was the ownership of discoveries. In 1790 the first Congress passed the U.S. Patent Act, which defined a patentable invention as novel, useful, and non-obvious to an expert in the field. Certain inventions clearly could be patented; the cotton gin, for example, or a transistor. But fitting biological discoveries into that definition was considerably trickier. Long and very expensive lawsuits have been fought to try to figure that out. Can someone own an antibody or a gene, for example? At the extreme, if someone discovered the key to life, could they patent it and own it? So determining what discoveries in the lab potentially had commercial value, and what actually could be patented was sometimes a complicated situation. Attorney George Gould had the responsibility for determining which discoveries made in the Institute's laboratories could be patented – and which of them had enough potential value so they should be patented. While initially Udenfriend had been wary of this potential interference in the pure research process, after the success of Spector's immunoassay he fully embraced the concept and became a strong advocate.

But patents were not mentioned in the original charter and never became a driving force inside the Institute. In fact, the back section of each year's annual report include a variety of information about the state of the Institute, including publications, awards and seminars. At no point though did it ever include patents applied for or obtained, which in most companies would have been an important means of measuring success.

George Gould had joined Roche in 1968, primarily to option potentially valuable technology from university labs to the company, then obtain a patent for it. "I filed patents for researchers all over the country, from Harvard and Johns Hopkins to Berkeley and Stanford. These places came to me and asked me to help devise licensing strategies for them. We were in the forefront with technology people in that area, even before the biotech start-ups came into existence." It was a job that required an understanding of both science and patent law. He had done post-graduate work in organic chemistry at NYU and had actually worked as a chemist at Merck for a brief time, but as he recalled, "I was working with a very obnoxious reaction at Merck and got concerned about my health – and the fact that I would stink when I went home. All my friends were studying to be lawyers and it seemed like a much more gentlemanly profession, so I went into law."

Molecular biology patent law was a perfect fit for him.

The difference in the way the Roche corporate research center and the Institute operated actually created problems for Gould; while scientists inside the Institute were permitted to publish papers on their research very quickly, scientists in the Roche research department often couldn't get their work into print for long periods of time. It was one of the few areas in which some minor resentment was fostered by the structure. While scientists inside Roche corporate research were trained and advised what to do when they had made a discovery, Institute scientists were not required to follow the same procedures. They didn't even have to disclose their results. "It was all based on personal interaction between Udenfriend and myself," Gould said. "He would monitor the research in all areas of the Institute. If he thought there was something of exceptional interest that would warrant at least considering filing a patent he would send it to me. I would find the right people in the right group to evaluate it and tell me whether or not it should be filed. If it was, Udenfriend would ask the scientist to briefly hold off submitting the manuscript for publication, giving us time to prepare and file an application."

When a patent was issued the inventor-scientist was asked to come up to the office of the Vice President for Legal affairs. "While Roche owned the patent," Herb Weissbach explains, "they paid the person who discovered the new intellectual property. The scientist would then be handed one dollar – in cash. That was it, one dollar, although at the end of the year there may have been a bonus and I suspect people were properly rewarded for their work. I suppose the dollar was paid for some legal reason."

In fact, it is difficult to determine precisely how many patents came out of the Institute. The best answer is, a lot, a whole lot; so many that in 1993 Sidney Pestka was inducted into the New Jersey Inventors Hall of Fame for work done in Nutley. The Institute itself owned no patents; any discoveries coming out of the building were owned by Hoffman –La Roche.

In addition, some of patentable discoveries coming out of the corporate research center were due to work initially done inside the Institute. As Gould explained, "After a brief time the scientists inside the regular Roche research groups started to be influenced by what was going on inside the Institute. Roche was looking for new methods for identifying and purifying the compounds that existed in nature. And there was a lot of interaction back and forth regarding peptides, proteins and purification of these materials. There was almost daily collaboration. People were walking back and forth, they were interacting. They were using all kinds of techniques but there was substantial cross-fertilization between the Institute people and the

239

research people. A lot came out of that and Roche probably was in the forefront of peptide and protein purification of any place in the world. There were several novel products that came out of that, and it's impossible to credit one place or the other."

Herb Weissbach believed firmly that among the many contributions the Institute made to the corporation was that "the focus of their research was markedly changed. In a period of 10 years Roche was transformed from a biopharma company, based primarily on chemical research, to a biopharma/biotech company, in fact a very large biotech company. Chemists still played a major role in this transition, and by 1980 the chemists were now involved in peptide and protein chemistry. I felt there was even a culture change in the company"

Gould became a familiar face in the Institute, keeping abreast of new work simply by walking around and asking people what they were doing. And while he was able to file numerous patents it is the one that got away that stuck forever in his memory. In the early 1970s scientist Alex Nussbaum, working with Daniel Luk, was conducting some interesting genetic experiments. As Gould remembers, "He had written a paper about putting together pieces of DNA. I got the manuscript before it was published. The bottom line, which I did understand, was that he was taking different types of DNA and putting them together; but I didn't fully comprehend how he was doing it. In fact, he was making what became known as sticky ends. He was the first person to be able to use restriction enzymes to put ends of different pieces of genes together. 'Gee,' I told him, 'That's a pretty fundamental thing and it sounds exciting.'

"He was much less excited than I was. 'Maybe,' he told me, 'but it's a technique that only going to be interesting to us genetic people. There's no real commercial value.'

"I spoke to other people but nobody was very interested. I couldn't get anyone else to buy into it so I let it go. If I listened to myself, I would have done the smart thing and filed a patent on it, and all of the recombinant DNA work would be patented by Roche for the next 17 years. It's impossible to guess how much that would have been worth. Impossible. That's the one I kicked around. But at the time nobody really believed there was any value to it."

While Gould considers that his greatest mistake, his biggest achievement was still a few years away. Within a decade the Institute and Genentech would combine to produce perhaps the greatest success in the Institute's history.

Meanwhile, the Institute continued to expand; by the mid-1970s the staff numbered more than 200 people, allowing

Udenfriend to report, "The Institute...has achieved its original goal of becoming a center of excellence for both research and training in molecular biology. This is evidenced not only by the substance of its research activities but also by the successes of the postdoctoral fellowship, graduate students and visiting scientist programs." Among those people who had joined the staff in recent years was Dr. Yasuhiro Furuichi, who was honored in 1975 as the Scientist of the Year by the Japanese Biochemical Society

The success of the Institute in training young scientists, then sending them off to other institutions and laboratories also was fulfilling the initial dreams. Syd Spector liked to remind people of a Chinese maxim, "If you plan for a year you plant rice, if you plan for decade you plant trees and if you plan for a millennium you teach." More than anything else, that belief was a fundamental goal of the Institute. Years earlier, when the concept of training scientists who would leave the Institute and might go on to make discoveries for competing companies had been explained to Mattia, he had expressed some reservations, but eventually had embraced the concept. In 1981, 10 years after the Institute building was completed there were 96 postdoctoral fellows and 9 graduate students being trained at the Institute. In addition, the initial resistance from universities, which were highly suspicious of a major pharmaceutical corporation's motives, had long been satisfied and 19 staff members held adjunct faculty appointments at those universities. Included in that group were Udenfriend, Sidney Pestka, as well as both Weissbach brothers, all of whom were serving as Adjunct Professors in the Department of Human Genetics and Development at Columbia University's College of Physicians and Surgeons.

The number of publications also had increased annually. In 1975 alone Institute members reported 173 publications and 58 abstracts, and the number of seminars -- in which staff members as well as invited speakers presented their research or spoke on topics of interest to a limited audience – were steadily increasing. In 1976, for example, there were eighty departmental seminars and an additional twenty Institute seminars – a substantial number, especially at a time when real dollar values were being attached to biochemical discoveries, increasing the value of this type of completely open scientific information. At a time when the world of science was merging with business, and start-ups were blossoming, the openness of the Institute about its research and accomplishments was going against the trend.

As Herb Weissbach reported, "At the end of our first decade many of the initial goals that Mattia envisioned had been reached and the strengths of the individual departments are now quite evident."

243

At the celebration marking the end of the Institute's first decade speaker Arthur Kornberg remarked, "The years ago when Sidney Udenfriend told me about the plans for the Roche Institute of Molecular Biology and its remarkable Charter for basic research, I told him that the Institute as an appendage of a major pharmaceutical company, couldn't possibly retain this character and integrity for very long. After all, a company is under constraint to report quarterly profit and loss statements whereas the practical benefits from basic research are not likely to be seen for decades. Now, ten years later, I must quote from Winston Churchill who said, 'It is a most nourishing and stimulating diet to eat one's own words.' And that is what I will proceed to do.

By the end of those first ten years the Institute's three departments had clearly delineated their areas of interest. Herb Weissbach's Department of Biochemistry continued to focus on protein synthesis and gene regulation, developmental biology and molecular aspects of membrane transport. Arthur Weissbach's Department of Cell Biology, which initially had been called the Department of Microbiology had become a mixture of research interests which included cell biology, developmental biology and its original focus, virology. In 1977 the name of Sidney Udenfried's Department of Physiological Chemistry was changed to the more accurate, Department of Physiological Chemistry and

Pharamacology to represent, as Udenfriend wrote, the reality that this was the most diverse department and "its interests comprise the studies of at least nine independent investigators." And while Udenfriend remained Director of the Institute he was replaced as head of this department by Syd Spector.

A new position, Laboratory Head, was created at the end of the first decade to recognize those scientists whose exceptional achievements had earned special recognition, and the first men to receive that honor were Aaron Shatkin, Ron Kaback and Bernard Horecker.

Among the most significant research done at the Institute in these early years was Aaron Shatkin's discovery of a unique cap structure in the 5′-end of reovirus mRNAs, which explained how cells recognized the beginning of the messenger RNA and initiated the protein synthesis process. – His discovery, which was titled " Structure and Function of Methylated Nucleotides in Eukaryotic mRNAs "proved to be a vital step forward in understanding how RNA functioned. "During the past year," Shatkin wrote in 1975, "a wide variety of eukaryotic viral and cellular mRNAs were found to contain a blocked, methylated 5′-terminal structure of the general type m7G(5′)ppp(5′)N... This 'cap' structure is present..."

These caps, which he and his associate, Yasuhiro Furuichi discovered were found on all nuclear-transcribed cellular mRNAs, as well as mRNAs of many viruses, proved to be involved in almost every aspect of mRNA metabolism in animal cells. While initially its importance was not widely understood, eventually, as scientific historians noted, "The history of the discovery of the presence of 5' caps on messenger RNA and its subsequent scientific flowering is a good example of how a seemingly small biochemical observation can lead to the deeper understanding of a wide range of fundamental molecular biological phenomena."

In addition to discovering the existence of the cap structure, Shatkin and Furucihi's experiments demonstrated its importance in the initiation of the translation process- how proteins are synthesized, but also its role in stabilizing the messenger RNA structure. It was a large stepping stone towards gaining a complete understanding of the genetic processes.

Arthur Weissbach believed that Shatkin's research should have been awarded a Nobel Prize. Several years later though, his work would enable him to become the first Institute scientist to be elected to the National Academy of Science for work done at the Institute, and his discovery was acknowledged with his being named Head of the Laboratory of Molecular Virology. Shatkin was precisely the type of scientists Art Weissbach was referring to when he talked about having a passion for science. Science was his world and he enjoyed sharing it with as many people as possible. A post-doc who worked with him later in his career remembers a day he had been working with one of Shatkin's post-docs, and after helping him complete an immuno-fluorescent staining he suddenly heard Shatkin shouting across the entire room, "Hello Doctor, I liked your staining. It is beautiful."

Nat Brot describes Shatkin as the proverbial "gentleman and a scholar." One of the points that made the Institute function so well is that the founding scientists, including Shatkin, established a collegial and welcoming atmosphere. Rather than walls, the feeling was inclusiveness. "He was warm, modest, fun-loving and caring," Brot remembers. "But his personality was surpassed only by his scientific contributions."

"Starting from the janitor any person in the lab was so at ease talking to him," Amiya Banerjee said. "He would put his hand on your shoulder and ask, 'Hey, how you doing?' But he meant it. He was my friend, philosopher and guide at every stage of my career at Roche. When I became an American citizen he was the witness. When my wife and I had a baby, Aaron and his wife Joan were the first people to come to our home.

"But he was always so much full of science. Whenever he would meet one of his post docs the questions were always, 'What are you doing? Why aren't you doing this? Why aren't you doing that?' Question after question. And he kept in contact with every post-doc fellow who passed by his lab, which would be 60 or 70 people. They all called or wrote, and when I wanted to know where one of the guys in his lab was doing work and he would tell me where they were, what they were doing and how to get in touch with them."

Bernard Horecker, who had been elected to the National Academy of Science in 1961, also played a vitally important role in the success of the Institute. Like Ochoa, Horecker had been an early colleague of Kornberg's at NIH, where Horecker had been studying the effect of DDT on cockroaches; and while initially Kornberg had worked in Horecker's lab, eventually Horecker, Kornberg and Leon Heppel had set up a new section on enzymes in the NIH's Laboratory of Physiology. As Kornberg wrote of their time together during World War II, "I had hardly heard of genes. Genetics wasn't then considered in medical practice, or even biology, strange as it may seem. But I did have three colleagues with whom I could share my excitement (one of them being Horecker). We met every noon for lunch and reviewed the current literature very seriously, we taught ourselves modern

248

biochemistry. We read and examined the Journal of Biological Chemistry from cover to cover. We were close friends and mutually supportive." These luncheons became quite well known in science, and it was said the only day they did not meet was Christmas. While Kornberg went on to discover the first DNA polymerase, an enzyme that synthesizes DNA using the 4 chemicals A,C, G and T, Horecker made significant contributions to biochemistry by elucidating the pentose phosphate pathway, an important pathway in sugar metabolism.

In the early 1960's, while occupying the microbiology chair at NYU School of Medicine Horecker had written a summary of his life's work, "There now appears to be general acceptance of the view that the earliest forms of life arose in an environment which already contained most of the compounds which we recognize as components of the living cell..."

"Respiration, the utilization of oxygen for energy production, appears to have developed later as a modification of pre-existing fermentation mechanisms... Our problem is therefore to examine the known pathways of carbohydrate metabolism and to evaluate them in terms of their role in meeting the new requirements which arose as the organic components of the environment were depleted, one by one."

Art Weissbach was a postdoctoral fellow in his lab at the NIH. During that period Horecker had taken a three-month sabbatical to work in Melvin Calvin's lab at UC Berkeley, with the possibility he would be offered a Full Professorship. Both men were researching photosynthesis. Calvin was awarded the 1961 Nobel Prize in Chemistry, although there were many people in the scientific community who believed Horecker should have shared in that honor.

Eventually though Horecker did leave the NIH and became Chairman of the then-newly formed Department of Microbiology at the Albert Einstein School of Medicine. Horecker initially served on the Institutes Board of Scientific Advisors, but as his three-year term was ending he asked Udenfriend to join as an active member, and eventually was named Head of the Laboratory of Molecular Enzymology. "He was a tremendously hard worker," Weissbach remembers. "And he expected everyone in his lab to work just as hard as he did. For example, he had a rule that no one was allowed to read a newspaper in the laboratory. He really objected strongly when he found someone 'wasting time' reading the paper."

He fit nicely into the Institute; in addition to maintaining a highly productive laboratory he took special pride in his mentoring ability. Anna Skalka remembers him as "dapper and supportive of his students." Among them Akira Endo, whose research helped lead to the discovery of statins, and who recalls coming to America in the mid-1960s, where he "worked on the role of phospholipids in an enzyme system involved in the synthesis of bacterial cell wall lipopolysaccharides in the world-renowned department of a great biochemist, Bernard Horecker, at the Albert Einstein College of Medicine in New York City for 2 years, from 1966 to 1968."

When Ron Kaback was elected to the National Academy of Science in 1987 he was on a cruise with 500 tire salesmen, among them his father-in-law and brother-in-law. He remembers, "None of them knew what the National Academy was. When my brother-in-law announced I'd been elected, my father-in-law asked, 'Do you get free tickets to the Philadelphia Academy of Music?'"

Kaback was named Head of the Laboratory of Membrane Biochemistry; he had become fascinated by DNA membrane transport as a student at Albert Einstein Medical School in the late 1950's, a passion that he never lost. After hearing Arthur Kornberg describe work he'd done with Paul Berg, he became convinced "there must be another species of RNA, located in the membrane, that's involved in amino acid transport." He started conducting his own experiments, confident this had to be correct. In fact, at one point a mentor set up a meeting for him with Sir. Francis Crick. As he recalls, "I was ushered into the room, scared to death. But I told him about my work. Sir Francis listened patiently and, looking down his nose, said, 'No, no. That can't possibly be right.'

"I remember thinking to myself, 'What the hell does this guy know about transport?'" Crick eventually was proved to be correct, but Kaback embraced the lesson he had learned, "That's what science is all about. You come up with some idea, do some experiments, and eventually your idea turns out to be total baloney. But along the way, you're taken in a different direction that turns out to be much more important than the idea that originally got you into the whole thing."

Kaback's interest in membrane transport was, ironically, infectious. The noted Dutch molecular microbiologist Wilhelmus Nicolaas Konings once recalled attending a seminar conducted by Kaback while still at NIH. At that time he had just begun to develop cytoplasmic membrane vesicles from E. coli as a model system in which to study transport. His excitement about it was so striking that Konings dubbed the vesicles 'Kabackosomes,' the name by which they became known.

It was that passion and experience Kaback brought with him to Roche. He had learned, he later wrote, that "To be a good scientist, I think it helps to be smart—but not too smart. I think if you're too smart you may talk yourself out of too many experiments. And you can't be patient. Impatience is very important. You have to really want results. If you're too patient, nothing will ever get done."

Describing one of his reports, titled *Active Transport, From Membranes to Molecules*, Kaback explained, "The primary objective of this laboratory is to achieve an understanding of the active transport at the molecular level."

For Kaback, the Institute offered the compelling advantage of not having to search for grants to continue his work. Like so many scientists, he often railed against a system that required scientists to become dependent on outside financing. Years later, he said, "The act of writing grants is not only a waste of time, I think it's anti-intellectual. When you write a grant, you're not coming up with new ideas; you're trying to sell something to a study section. To me, that's not what science is about."

The reality of that struck Sidney Pestka when he left the Institute in 1986. Having spent his career first at the NIH and then at Roche he never had to write a grant to fund his research projects. Upon moving into a different world he found himself increasingly frustrated at the amount of time he had to spend writing and editing grant requests to secure funding.

The Institute was firmly established and had become a valuable intellectual and, somewhat unexpectedly, financial contributor to the corporation when CEO and President Robert Clarke retired in 1980 and was replaced by Irwin Lerner. Just as there had been concern when Clarke replaced Mattia, there was considerable trepidation that Lerner might want to make changes in what until then had been a very successful balancing act between the corporation and the Institute. Lerner's career track made people inside the Institute somewhat anxious: he had spent his two decades in the marketing division of the company, rising to Vice President. He had been the Director of Advertising when valium and several other extremely successful products were launched. And as he remembers, "When the Institute was launched a lot of us were scratching our heads trying to understand what it was all about. When we finally read a description of it, some of us cynics decided it was a big publicity scam. An Institute for purposes undefined? Not to find drugs, but to do whatever they wanted to do scientifically? This is madness, it's just throwing money away. Maybe it was great for building a public image, he thought, but like so many others I wondered what it was going to do for the bottom line."

There were reasons for concern. Roche had fulfilled the initial promises that Mattia had made to John Burns more than a decade earlier. There was nothing to prevent the new CEO from closing the Institute. And with the end of the valium patent in sight, meaning the loss of the exclusive rights to the corporation's most important product that clearly was a possibility.

Lerner too had been present at the creation of the biotechnology industry; he was the first drug industry member appointed to the newly formed Biotechnology Industry Organization. And throughout his career he had spent time learning about and understanding the science behind Roche's products. "At the beginning of my career in the industry," he said, "I didn't know a thing about pharmaceutical products. I got my first job by answering a want ad. Before the ad I bought a medical dictionary and learned how to pronounce all the difficult terms.

"I decided very early in my career I would get to know the scientists. I spent all kinds of hours in the laboratory, trying to learn about what made these products tick. Eventually I could describe every experiment they did, all the pharmacology that went into testing the clinical trials, where they were going and who was doing them. I was glued to the scientific establishment in the company."

But because he came from the purely commercial side of the business, he acknowledges, "A lot of people on the board got very nervous. All these hot shot scientific geniuses didn't know what I was going to do. Was I just going to drop the hammer on the whole thing and focus on commercialism or would I continue to support it. I could tell they were worried by the luncheons we would have. They sat there stiffly just waiting for me to say something nice about the Institute.

"But I was a big believer in the potential commercial value of biotechnology. I believed it was going to be a great new thing and I tried very hard to show them they had nothing to fear from me in terms of management decisions that might negatively impact the Institute. What had first turned me on to the Institute's value was the commercial development that came out of the Institute, not by direction but by serendipity. Abuscreen proved to be a very valuable product and I remember thinking, gees, these guys are doing something worthwhile after all, something we can make a lot of money on."

What may have surprised people was Lerner's commitment to research. His philosophy was simple, "You can't make money unless you produce useful, meaningful new drug products." To support that he began making substantial investments in research. "We put 23 cents out of every dollar back into research and development, which was pretty much unheard of in those days. In the non-pharmaceutical area most companies put about 5% into R and D. General Motors, for example, spent 5% of sales on research.

"In that research area doing well meant bringing in the very best and brightest people, trusting your research director and if he believes in a project, support it. Pump the money into it that's necessary to bring it to fruition." Lerner recognized that on the average only one of hundreds, or even thousands of compounds, is going to make it to the marketplace, but having seen what impact a single drug like valium could have on the company he was more than willing to continue making substantial investments in research. And after becoming CEO of the corporation he continued that commitment, becoming a frequent visitor to the Institute – as well as a strong supporter of this completely independent research unit.

Lerner's elevation was only one of several significant changes that impacted the Institute in the early years of the 1980s. In 1982 Sid Udenfriend stepped down as Director of the Institute and the natural choice as his successor was Herb Weissbach who had been the Associate Director from the beginning. "I was honored to be appointed his successor," Weissbach wrote, "a position that has special meaning to me since Sid and I have worked together for 30 years... My primary goal is to insure that the Roche Institute continues to be a center of scientific excellence."

Earlier that year Weissbach had become the second scientist at the Institute to be elected to membership in the National Academy of Sciences primarily for work done during both in his early years at the NIH and at Berkeley, but also for his research at the Institute. He had been involved in the discovery of the first coenzyme form of vitamin B_{12}, had done research in understanding the known metabolic inter-relationship between vitamin $B_{,12}$, folic acid and one carbon metabolism, and beginning with his studies with Nirenberg, and later at the Roche Institute, for his work in elucidating the role of the factors involved in synthesizing proteins on the ribosome.

Roche Members of the National Academy of Sciences. Left to right, Aaron Shatkin, Bernard Horecker, Allan Conney, John J. Burns, Sidney Udenfriend, Herbert Weissbach and Severe Ochoa. Picture taken in 1983 after Conney and Weissbach had been elected. In 1987, Ronald Kaback (not shown) was elected to the Academy. Reprinted from Weissbach,H. Reflections on the Roche Institute of Molecular Biology after 20 years. In Research at Roche, Reminiscences and Reflections for Fritz Gerber (1989), editors, Jurgen Drews and Fritz Melchers, Roche Basle

Herb Weissbach had barely gotten comfortable in his new position when John Burns announced that he intended to retire. Since the founding of the Institute in 1967 Dr. John Burns' areas of responsibility had changed several times as the corporation restructured; but through them all he had remained the corporate representative to the Institute. The creation and the success of RIMB ranked very high among his accomplishments. As Richard Snyder pointed out, "John was a master at problem solving and bringing people together. There always was a divide between the corporation and the Institute and he did a remarkable job avoiding the pitfalls and leveling the ground. He greatly encouraged scientific interplay between corporate R&D people and the Institute and deserves credit for nurturing and protecting the Institute in those early years until it was strong enough to exist on its own reputation."

The question became who would replace him. The two most likely candidates within Roche, were Herb Weissbach and Ron Kuntzman; Kuntzman had started at Roche as a scientist and through the years had become a Vice President and Director of Corporate Research, as well as a close friend and key advisor to Burns. When Burns elected to retire in 1983 Lerner asked both Kuntzman and Herb Weissbach to write proposals outlining the changes they might make if they were selected to replace him.

While there was some professional competition between Kuntzman and Weissbach, they also were good friends, having met at the NIH in 1955 when Kuntzman was hired as a graduate student. In fact, Kuntzman initially was trained by Weissbach.

Herb Weissbach submitted a more radical plan for future research at Roche, stressing a rapid move into the new fields of biology. That might have been bad timing. Probably because Kuntzman had spent his career in the corporate culture and had an excellent reputation at Roche as an administrator, he was selected. It made little difference; For a period of several years, until Ron Kuntzman retired, the two friends worked well together.

So the comparatively new leadership of Irwin Lerner, Ron Kuntzman and Herb Weissbach was in place when the Institute faced the biggest challenge to its continued existence. In 1983 the valium patent ran out, allowing competitors to produce and sell their versions of the drug. The copycats, generics, were coming onto the market, substantially reducing Roche profits. To deal with that reality, Lerner had to resort to layoffs. Operation Turnabout, it was called, and it put the existence of the Institute in jeopardy. As many as 10% of all Roche employees would be asked to leave. The question became which division would be hardest hit? And would the Institute that Udenfriend, Burns and the Weissbach brothers had built continue to exist in that form?

Valium had been largely responsible for Roche's spectacular success. Kuntzman remembers attending a budget meeting when he joined the company in the late 1960s. "During the meeting someone asked the budget director if we were going to have an increased budget in the next year. The answer was 'Increase your budget by 10%.' Every year that was the answer, 'Increase your budget by 10%. This went on for years and years as the valium money just kept coming in. But the year after valium lost its protection we had a 10% reduction. It was very difficult, something no one had dealt with before and there was a lot of fear where those deductions were going to come from."

The practical reality that people were going to have to be fired actually reinforced a recommendation that Herb Weissbach had received from the Scientific Advisory Board. Unlike many corporate boards that don't actively participate in that corporation's management, the Institute's Scientific Advisory Board took its role seriously. Their function was to act as advisors to the Institute and also to advise Roche management about the Institute. Based on its recommendations the Director and Associate Director would get feedback from corporate management. As Weissbach was taking over from Udenfriend the advisory board made a curious finding: The Institute from the start had done too good a job. They told Weissbach that the

261

Institute had too many tenured scientists. "No one is leaving," they reported. They either had to be encouraged to leave or let go, the board warned, "You have to get more turnover or senescence is going to set in pretty soon."

Weissbach recognized the wisdom of that suggestion, writing artfully in 1983, "The problems we face are quite different than those that had to be solved in the fledgling years of the late 1960s and early 1970s. The Institute has reached a stage where there is a tendency for less turnover, as a greater proportion of the staff settle in to permanent positions. This stability offers many advantages but, at the same time, it is clear that we must have sufficient flexibility to change with the times. The infusion of new technologies is inherent to the success of any research organization... We must not lose sight of this and one of my most important responsibilities is to insure that the Institute retains its vitality in the years to come."

So Herb Weissbach already was dealing with the problem of finding a way to bring renewal and reinvigoration to the Institute when Operation Turnabout was announced. He knew he had to protect his senior staff, the "heart of the Institute." But he also was certain that his staff was so talented that they would easily find academic positions if the time came that the existence of the Institute was threatened.

It was a difficult situation for Lerner, and his professed dedication to research was about to be severely tested. Inside the Institute he was perceived to be more of a businessman than a scientist, and the fear was that research, one of the most significant budget items, would take a substantial hit. And if money had to be cut from the research budget it was far more likely to be taken from the Institute, whose contributions to the bottom line weren't always easy to calculate, than from the profit producing corporate research center.

Lerner initiated Operation Turnabout by bringing in the consulting firm McKinsey to make the initial recommendations. He tried to make what certainly was a difficult time as easy as possible by keeping employees as informed as possible about the impending layoffs. Eventually his "fundamental, unrelenting belief in the power of research and development to produce meaningful results," softened the impact on the Institute.

"The Institute did get hit," Ronald Kuntzman admitted, "but we were able to limit the impact. I knew management was going to protect the scientists at the Institute as much as possible, so I used that to save some jobs. I knew McKinsey would look at the bottom line, that's what they were supposed to do, but often important research contributions don't show up there.

"One of the moves we were able to make was moving Dr. Allan Conney, the Head of the corporate Laboratory of Experimental Carcinogenesis and Metabolism, and his people into the Institute. Conney was doing important basic research but I knew it wasn't going to look good so we had to get them out of there. Herb Weissbach did some contortions to make it work, but we were able to save their jobs."

Weissbach remembers it well, pointing out, "Ron Kuntzman was insistent that I had to take some of the groups from the company. He did ask me, but he made it clear this had to be done. Although scientifically it turned out to be a really good move, as the budget did not allow us to hire any new people at that time, this marked the very first intrusion from the company in how the Institute was being run. It was the first time that barrier had been crossed, and psychologically we saw the company forcing the Institute to do something. Inside the Institute and that raised the fear that the company was now going to start directing research done there. It was the very first indication that things were going to change."

To make everything fit, Dr. Anna Skalka and her assistants moved back into the Institute from the corporate side to become Head of the newly formed Department of Molecular Oncology. Allan Conney and his people joined that group, "bringing together," as Weissbach explained, "a wealth of scientific talent in new and rapidly moving areas."

It was a difficult and challenging period. Weissbach presented the move to the Institute staff that "both the Conney group, and Ann's group were fine fits for the Institute and expanded its scientific strength." What helped was that Alan Conney had been elected to the National Academy of Sciences in 1982 and everyone had a high respect for Ann Skalka. However, it took a couple of years to reassure the staff that the Institute charter remained intact. Not a single senior scientist lost their position during Operation Turnabout; in fact the additions actually increased the size of the senior staff. However, during the next few years several of the most talented scientists, people who had joined the Institute during the very early years, decided to move to academic institutions.

Eventually the necessary cuts were made, and the strategy enabled Roche and the Institute to maintain its position as one of the world's leading pharmaceutical companies – that and arguably to make the institute's most important discovery.

Chapter Six

How the Institute Changed Hoffmann-La Roche Forever

It is probable that Sidney Pestka's discovery of interferon was the most significant achievement made in the entire history of the Roche Institute, for both the actual product as well as how it fundamentally changed the entire company. It was interferon that brought Roche into the new world of biotechnology and forged its relationship with Genentech.

At one point in the early 1980s, attorney George Gould remembers, "There was a lot of ambivalence about the future of biotechnology at Roche. A lot of knowledgeable people in marketing did not think there was a future in protein drugs and wanted to focus on small molecules. They wanted to use our biotech capabilities as a means to screen for new small molecule drugs and move away from pursuing proteins as potential therapeutics."

It was the possibilities offered by interferon that prevented that from happening. Had those people swayed Irwin Lerner, Roche arguably would not have remained a leader of the industry.

Like most of the scientists in the original group, Pestka came to the Institute from the NIH, where as a post-doc he had worked with Marshal Nirenburg helping to crack the genetic code. Later, while working in the NIH's National Cancer Institute, and still studying ribosome function in protein synthesis, Pestka first became fascinated by the research being done by Hilton Levy and Samuel Baron, who were trying to unlock mysteries of this strange protein, called interferon. Long before its discovery in 1957 by Alick Isaacs and Jean Lindenmann at London's National Institute for Medical Research it was known that in cultures certain viruses could block, or interfere, with the actions of a second virus. As it was described at the time, "one virus can interfere with the growth of a number of unrelated viruses." Researchers simply did not believe that even an inactive virus was capable of preventing an infection. The problem was that no one knew how or why this happened – until Isaacs and Lindenmann demonstrated that the growth of an influenza virus in chicken embryos could be prevented by injection of a heat inactivated influenza virus. The heated inactive virus was stimulating the production of a protein in the chick embryo tissue that was interfering with the growth of the influenza virus. "To distinguish it from the heated influenza virus," they wrote, "we have called the newly released interfering agent, 'interferon.' The meaning, and the possibility, were incredible: This protein might be used to prevent or fight – and defeat – viral attacks. It was thought to be one of the key weapons in the body's natural immune system. If that was true, and it could be isolated, it might help scientists understand how the human immune system functioned – and perhaps even how to enlist it in the fight against a plethora of debilitating and even fatal conditions and diseases. That potential got a lot of people excited. Interferon. It had a name and scientists had seen its capabilities. But beyond the fact that it existed almost nothing about it was known. Among the many problems involved with figuring out how it worked was that it was extraordinarily difficult to isolate; because the body naturally produces it in such infinitesimally small quantities collecting a sufficient amount to use for experiments was both time consuming and expensive. To solve that initial problem several of the large pharmaceutical companies began working on interferon inducers, compounds that would induce the cells to produce their own endogenous interferon in large quantities to fight off a perceived viral attack. Scientist David Goeddel, one of the very first people recruited for Genenetch by Bob Swanson, explained, "Interferon is just produced in minute amounts in response to viral infection, and no one had been able to make a lot of it in pure form, and sequence it and study it." Mostly there were a lot of reports that partially purified and small amounts of it seemed to show intriguing potential as a therapeutic.

It proved to be an extremely difficult project, as the compounds (inducers) turned out to be too toxic, too variable and simply too difficult to use. And until that could be solved it remained very difficult to conduct the type of experiments that needed to be done. The route to discovery means navigating through failures, but the lack of interferon to work with made progress excruciatingly slow. Through the 1960s much of the research and the production of interferon was taking place in Finland, but as biotech start-ups proliferated it became a popular research project. A lot of people believed Interferon had the potential for changing the world and spent large amounts pursuing it. "So little was known about it, Goeddel pointed out. "Scientifically it was a super-challenge to find the gene and make it. It was just about the ultimate project. It had the chance to be a transformational product – that would make the people who figured it out very rich, too."

Interferon had become Sidney Pestka's passion. Herb Weissbach had known Pestka since they were both working on protein synthesis in Marshall Nirenberg's lab and later they had been close friends and neighbors living in the same Rockville, Maryland community. At times Sid and Joan Pestka would babysit the Weissbach's children. So when Weissbach decided to build a strong protein synthesis group at the Roche Institute Pestka was a natural person to recruit. Like most of the scientists who accepted offers to come and continue their research at the new Institute, it was the lure of being able to do completely supported experimentation that led Pestka to become one of its original members. His fascination with the way the world works had begun at a very young age, soon after his family had moved here from Poland. As he once explained, "It was stimulating to see chemicals change the color of fluids, to construct crystal radios, and to make caramel from sugar—however, my mother's pots and pans were never the same afterward. It seemed that I constantly thought about new ideas to implement."

Pestka was an exacting and serious scientist – who looked somewhat like Woody Allen. In fact, a story told inside the Institute exemplified that; one day he and his wife were in New York City when he was stopped by a young man. 'Hey! Woody!' he said. 'Can I have your autograph?' Pestka smiled and took the man's pen and paper and signed the autograph. And as he and his wife walked away the man asked, 'Who's Sid Pestka?'

During the Institute's first year, while the building was under construction, Pestka continued working at the National Cancer Institute. Among the reported experiments in the Institutes first annual report was Control Mechanism in Normal and Cancer Cells, authored by Pestka and Robert Gallo, who later would become internationally renowned as the co-discover of the AIDs virus. In that report Pestka wrote, "Control of cellular growth, differentiation and carcinogenic transformation may be mediated by regulation of protein synthesis."

The search was on.

Pestka played an active role in the Institute during those formative years. At the very beginning Udenfriend established an Institute Senate, consisting of the entire permanent senior staff, to advise him as policies were established and the building was designed and built. For the first year Dr. John Hutton served as this Senate's first chairman – but at the end of his year-long term he was replaced by Pestka.

Progress towards completely understanding the role of interferon in the cell was slow. Very little was known about its chemical structure or its composition. But what was known was tantalizing. The Finnish Blood Bank undertook a large scale program to provide interferon for researchers around the world.

They were able to produce a small sample from white blood cells taken from 90,000 units of human blood. This was natural interferon. It was given to four patients suffering from Hepatitis B, a dangerous and sometimes fatal disease. Three of those patients responded, two of them were considered free of the disease. It was a remarkable display of the potential of interferon. But it remained too difficult to obtain: All of the interferon collected from 90,000 units of blood could be used to treat, at most, 250 Hepatitis B patients. The human body simply did not produce enough interferon to make a viable product. It would be like trying to satisfy the world's need for milk from a small herd of cows. Sources other than human white blood cells had to be found if interferon was ever to fulfill its promise.

At about the same time Sol Spiegelman, a Professor at Columbia University's Medical School as well as the Director of the Institute of Cancer Research was researching the perplexing possibility that certain forms of cancer might be caused by viruses. That possibility made understanding interferon even more exciting and important.

Pestka had decided when he came to RIMB that he would focus on Interferon. He continued his work on protein synthesis as he slowly moved his research into studying interferon. The Biochemistry department had other scientists interested in

272

protein synthesis at that time, including Nat Brot and Jim Ofengand. As the field developed Weissbach was approached by Academic Press asking if he would put together a book on the subject. He asked Sid Pestka to co-edit the book which was titled Molecular Mechanisms of Protein Biosynthesis and published in 1977. It was the first book published describing in detail the individual reactions involved in making a protein in the cell, i.e., how the information in the messenger RNA is translated.

Initially Pestka attempted to purify rabbit Interferon, but for some inexplicable reason whatever activity he was able to generate always disappeared. One of the hallmarks of a great scientist is that failure, rather than dampening enthusiasm, sparks curiosity. And curiosity is the first step in discovery. Rather than giving up in frustration, Pestka simply worked harder.

Obviously this was not the only research being done in his lab, and he did have other successes: for example, for the first time the immunoglobulin heavy chain mRNA was isolated in his lab.

Working with a large amount of cell extracts provided to him by NYU microbiologist Jan Vilcek, Pestka successfully isolated two distinct types of interferon. He reported the result of his experiment in RIMB's 1976 annual report, the first mention of interferon research. "Interferon was characterized as an antiviral entity in 1957," he reported. "Viruses, double-stranded RNA and other compounds have been found to stimulate interferon synthesis by cells. At present, however, surprisingly relatively little is known about interferon induction, biosynthesis, action or structure... Recent clinical trials indicate interferon can be used to treat chronic hepatitis... but because of the difficulty in obtaining large amounts of human interferon, clinical trials have been somewhat limited."

And then the details of the experiment, "To study interferon biosynthesis, we have partially purified interferon messenger RNA from induced human fibroblasts and have translated the mRNA in various systems with the resultant production of biologically active human interferon... The synthesis of interferon represents the first biologically active eukaryotic (a living cell with a nucleus) protein synthesized in a cell-free extract. The ability to assay interferon mRNA should enable us to purify it; and the ability to synthesize biologically active human interferon should enable us to delineate its biosynthesis and elusive structure."

It was a substantial step forward. They were learning more every day about this elusive molecule. And other labs inside the Institute also contributed to the research effort. As the recombinant DNA techniques, pioneered in 1 in the laboratories of Boyer and Cohen, became more available Pestka realized he could clone the cDNA. This really was cutting edge biotechnology, and the race was on to solve this puzzle. In her lab one flight up in the Institute, Anna Skalka was working with a bacteriophage that was a very important vector for cloning genes of higher organisms, "And so," she explains, "I was in a very good position to enter the field of recombinant DNA technology, using bacteriophage and plasmids to study the viruses of higher organisms. Sid Pestka got interested in cloning the interferon gene but the people in his lab knew nothing about cloning.

"So Pestka sent up one of his post-docs to work with me and we taught him how to clone, and eventually they were able to clone the interferon gene. Basically, cloning means taking a piece of DNA, such as a plasmid or a virus, that will grow in bacteria or in a cell. You begin by cutting it open chemically and inserting another piece of DNA. Then you put it inside a bacterium or cell and it replicates. It makes many copies of itself -- including that particular gene that you inserted. And then you can take those many copies of the gene and put it in the particular plasmid in which it would be expressed and that gene expresses itself and makes the protein. You can make bucket loads of it. That's what Pestka wanted to do; he stuck a human interferon gene into my Lambda bacteriophage, put that back into bacteria and grew up the bacteria. It was all Pestka's baby: he was smart enough to know what he wanted to do, get the tools he needed and do it. That was the origin of the first product produced by Roche from the new recombinant DNA technology." But the ability to produce pure interferon was still a long way from reality.

Udenfriend had assigned two very accomplished young scientists, Dr. Stanley Stein, who had received his doctorate several years earlier at City College and Menachem Rubinstein, a post-doc fellow from Israel's Weizmann Institute of Science, to assist Pestka on his interferon purification project. Rubinstein had been attracted to the Institute by its growing international reputation, made possible by its accomplished roster of scientists.

In order to be able to clone the interferon gene it first had to be purified; in the two decades since its discovery so many scientists had wrestled unsuccessfully with this problem that it had become known as 'misinterpreton.'

As Pestka and Joe Tarnowski wrote, "Interferon used in experiments as well as initial human trials was essentially a crude protein fraction in which less than one percent by weight consisted of interferon. Because of the use of such crude interferon-containing material, it was not clear what activities of these preparations were, indeed due inherently to the interferon present and what activities were due to the numerous other contaminating proteins... without essentially pure interferon it was not possible to demonstrate definitively whether or not a particular activity was due to the interferon molecule itself." Additionally, little was known chemically about interferon and successfully isolating a sample would allow researchers to both determine its composition and structure, and from the amino acid sequence be able to identify the interferon gene.

When giving advice to young scientists, the very first thing Pestka emphasized was perseverance; pointing out the necessity of focusing on important problems while minimizing the time and effort spent on trivial problems – because the rewards are far greater. According to the Weizmann Institute newsletter, Pestka and his team were almost obsessed with solving this major problem. "Rubinstein worked day and night on the job using techniques he had developed with his mentor, Sid Udenfriend at Roche, and he recruited his wife, Sara to work as a lab technician (in the Institute) while they split the care of their two young children. He recalls that a part of the solution to isolating interferon came to him in a dream: When he went to the lab (in Nutley) the next day he found his dream, indeed, had contained the insight he needed to make the necessary breakthrough."

The eureka moment, like almost all great discoveries, was the product of countless small steps and innumerable successful and failed experiments, and the result was not what had been anticipated. One of their first attempts was to purify interferon extracted from human white blood cells, or leukocytes as they are known. That experiment stalled when the interferon got stuck on the liquid chromatography column, an instrument used to separate, identify and quantify the components in a mixture. Normally either methanol or ethanol would elute, or release, the substance but in this case for some inexplicable reason neither one worked. Pestka suggested they try to use propanol, which surprised both men, as it was well known that proteins could lose activity in organic solvents. In a different lab they might have pointed that out to the principal investigator. But that's the way Pestka ran his lab: There was a rule in his lab that when a lab assistant for any reason decided not to follow his suggestion Pestka would either tell someone else to do it or do it himself.

But one way or the other it would get done his way – which did not serve to strengthen the relationship between Pestka and his assistants. So they did it his way – and were stunned when they eluted the column with propanol and the interferon came right off. Eventually they found that the leukocyte interferon could be separated into 10 closely related but different proteins. While the differences between them were slight, further research confirmed that the antiviral or growth inhibiting activities actually were carried out by different forms of interferon.

It was clear that Pestka had isolated the human interferon gene for the first time. But the real surprise was that rather than a single interferon there appeared to be 'multiple species' of interferon. That made figuring out how to purify large amounts of each species even more difficult. On the other hand the situation was ripe for recombinant DNA technology. He had isolated the gene of one of the species and successfully put it into bacteria, but now he had to figure out how to program that bacteria to make the protein. While Pestka had successfully cloned the gene in bacteria, he hadn't figured out what regulatory signal was needed to tell the bacteria to start producing the protein. "He could put the gene into bacteria, but he couldn't get the bacteria to make the protein," Herb Weissbach remembers with frustration. "None of us at that time knew how to make that bacteria read the interferon gene as compared to reading its own genes."

Human cells are programmed so that specific genes are used in specific tissue or organs inside the body. Different places in the human body require specific genes turned on to grow and function, and there are mechanisms to insure that happens. It's why our organs are located where they are and do what they do. There is a region of the DNA associated with the gene, think of it as a switch that only is turned on when it is in the right place. Essentially, that gene is saying, "I know my function. To accomplish that I have to be in the right place to accomplish my job. There are specific signals in mammalian cells that tell a specific cell, go ahead and transcribe this DNA into a mRNA and translate the mRNA and make the protein. For example, if I'm programmed to make growth hormone I only want to be turned on in the pituitary gland, where growth hormone is made and insulin is only turned on by the insulin gene in the pancreas, even though all the cells in the body have the complete set of 20,000 genes." Until a specific signal activates a gene, like flipping on a light switch, the gene refuses to let its information be used to make a protein. "If you want to take a plant gene, for example, and put it into a bacteria to make the plant protein, you have to figure out how to properly add, or construct the plant gene in the DNA so it has a regulatory region (switch) next to the gene that the bacteria will recognize. Otherwise you will never be able to turn the gene on.

Pestka had successfully cloned the interferon gene. He just didn't know how to flip the switch. "We just weren't smart enough at that time," Herb Weissbach says. "To move a gene from one species to another you have to make sure that not only is the coding correct for the protein but that you know the proper signal to turn that gene on in the new host. The host, at that time usually bacteria, is not going to express it, turn it on, unless it can read it properly, which means it needs to have the right regulatory sequence next to the gene."

They were tantalizing close, yet still far away from success. While Pestka was struggling to solve this problem Herb Weissbach was giving a presentation about his protein synthesis studies at a conference held in 1980. Among the other people attending that conference were a small group of scientists from the new biotech start-up, Genentech. After the meeting, Weissbach remembers, he was approached by them and "they were very candid. They asked outright, 'We heard you might have the Interferon gene.' I sort of hemmed and hawed. Finally they told me, 'We'd like to come to Roche and talk to you about making human interferon as a drug.' They offered to sign a confidentiality agreement. I said I would convey their message and try to arrange that. As soon as I got back to Nutley I discussed what happened with John Burns. We agreed that it made sense. This turned out to be the beginning of an exciting relationship between the two companies that had a profound impact on both companies.

Several weeks later Herb Boyer and Bob Swanson came to the Institute. The meeting did not start well; Swanson's flight from San Francisco had been delayed and by the time he reached New

Jersey his reserved room at the Ramada Inn had been given away. "It was midnight and they had no rooms left," George Gould says. "So they set up a cot in some sort of utility room and let him sleep there. The poor guy was dead tired. And by the time he got to the Roche campus he was really peeved. What kind of company were we that we couldn't get him a room in a hotel? Our relationship did not begin on a very good framework."

But the actual meetings went well and things moved rapidly. They were told that Pestka had cloned the alpha interferon gene but that the expression level was very low. In response Boyer pointed out that they had successfully made human growth hormone and human insulin in bacteria and knew how to do it. They suggested some sort of undefined partnership.

After that meeting Herb Weissbach met with Lerner, who asked, "How long is it going to take you guys to find out how to do it?"

"It's a difficult problem," Weissbach responded.

"I understand that," Lerner agreed. "But how long is it going to take you to solve it?" How long, he was asking, until Pestka could identify the regulatory region in the bacteria and turn it on?

"I don't know," Weissbach said, perhaps even a little optimistically. "It might take six weeks. It might take six months, even a year."

While no one knew how – or even if -- interferon would benefit Roche, waiting patiently was not the way business was done in the pharmaceutical industry. Being one day late to patent a discovery literally could cost billions of dollars. At about the same time this was happening, for example, scientist Charles Weissmann had co-founded the Geneva, Switzerland-based biotech start-up Biogen, and was rumored to be making important breakthroughs in isolating and cloning interferon. Supposedly as many as four other companies also had interferon projects in progress.

Lerner frowned. Even if interferon was not the magic compound people believed it might be, it certainly made sense to move forward with the interferon project as fast as possible. No one had any idea how close one of Roche's competitors was to being able to successfully clone interferon, but Lerner knew the company could not afford to wait. This was the first real opportunity for large pharma to produce a biotech product created by genetic engineering. The stakes had been raised even higher in 1978 when the University of California filed the first application for a patent on a gene; in that case the gene that

283

produces human growth hormone. That filing introduced the vitally important concept that genes could be patented, or owned, by a company, adding even greater financial incentive for the biotech world.

Lerner told Herb Weissbach, "Go out to Genentech and find out if they really can do this." And with that, the Institute took the next important step towards forming the relationship between the two companies that would fundamentally change the entire company and the industry.

Genentech at that point was still a relatively small start-up company. But apparently scientist David Goeddel had solved that elusive problem; he had figured out how to construct a human gene with the right bacterial signal or regulatory region in front of it, so that the bacteria would turn the human gene on and make the human protein. With this extraordinary technology Genentech had pursued the low-hanging biotech fruit, successfully cloning human insulin and human growth hormone, knowing that a substantial and profitable market existed for both of them. Until then physicians had been using pig insulin to treat diabetes, and they were isolating growth hormone from pituitary tissue to treat people of short stature.

The success in producing insulin marked a major milestone in biotech history; and it was a moment Goeddel will never forget, "I'd worked the entire night. I was in the lab with Art Riggs, who was a collaborator on the program. Art came to work at his normal time, about 9 a.m. and happened to walk into the lab just as I was getting the data up. I said to him normally, 'We have it. Here it is.' It was a magical moment. It was clear we had made a tiny amount of insulin."

Producing HGH was a far more complicated process, "There were still a lot of unknowns," Goeddel continued. "We didn't know if we could make the mature protein, which was the key. But eventually we had assembled a semi-synthetic gene. It was part natural, part-synthetic, but it coded for human growth hormone. We inserted it in an E. coli bacteria which allowed it to produce mature HGH. That was a big breakthrough. It told us that we were going to be able to do a lot more with this technology. To me, that success said, 'Hey, this technology is going to be applicable to all sorts of things. This whole industry is going to be very successful.'"

Coincidently, scientists inside Roche previously had considered attempting to clone human growth hormone. Roche had licensed a patent from Berkeley scientist C.H. Li, who had successfully synthesized it. And there was considerable debate whether the company should do the recombinant work necessary to isolate a gene for HGH. Eventually the decision was made not to spend the time and money to do so because the existing small market for it, which essentially consisted of children with dwarfism, was being satisfied with growth hormone taken from the pituitary glands of cadavers. But at just about the same time Genentech successfully cloned the gene, it was discovered that cadaver HGH in some cases carried with it the potential for a fatal disorder disease, caused by prions, known as Creutzfeldt-Jakob disease. Almost overnight a market for Genentech's safe and effective product came into existence and quickly became a very profitable product for them.

Growth hormone was a big step forward in recombinant DNA technology. As the magazine New Scientist announced in 1979, "Human Growth Hormone (HGH) is the most complex protein so far produced using recombinant DNA technology, and its synthesis involves several breakthroughs in genetic engineering techniques. These advances demonstrate that still more complex substances, previously beyond the scope of synthetic DNA techniques can efficiently be produced...

"Adding a genetic switch which allows the gene to reproduce in a foreign medium... in what they see as a major advance, the Genentech researchers have been able to isolate the switch used by the bacteria to turn on one of its genes, the β galactosidase gene. This switch makes the bacterial production of the hormone much more efficient, increasing by a factor of 30 the quantity obtained." And this switch also could turn on a variety of human genes.

The article continued, "If Genentech can make HGH, what else can it make? One of the substances Genentech is believed to be working on is Interferon, which inhibits many viruses. Interferon is seen by many as having the potential to become the first broad-spectrum anti-virus drug. As such, it would hold a position similar to the first antibiotics in its significances for the treatment of illnesses... At present, Interferon is available only from human cells and is expensive.

"Despite all the problems, interferon will be genetically produced by someone before the end of 1980, predicts Bob Byrnes, Vice-President for Marketing at Genentech, but he is not prepared to predict who will win this race. More than a dozen industrial laboratories are working on recombinant DNA technology in the U.S."

Weissbach flew to San Francisco with Roche patent attorney George Gould. It was clear that both companies could benefit by forming a relationship. While Roche needed Genentech's cutting edge technology to actually express the interferon gene, Genentech needed the clone from Roche to turn a laboratory discovery into a successful product. Rather than marketing its synthetic human insulin, for example, Genentech had licensed the production to Eli Lilly. What they had in scientific know-how they lacked in production, marketing, the entire FDA approval system and everything else required to get a product into the hands of pharmacists.

"We met with Swanson, Boyer and Dave Goeddell in a small room at Genentech," Herb Weissbach recalls. I believe there was a 250 ml. bottle filled with white powder sitting on the table. To me, it looked like it was filled with sugar, and that was reinforced when we were asked if we wanted coffee. It wasn't sugar; Swanson picked it up and said something to the effect that, sorry, it wasn't sugar in that bottle, it was human growth hormone that had been made in *E.coli* bacteria. I just stared at that bottle. I was stunned; there was more human growth hormone in that little bottle than had ever been made by any other technique. I was looking at evidence that biotechnology could change the world."

I spoke to Dave Goeddell that morning and I was so impressed by what he had accomplished. I whispered to George Gould, "We can go back to Nutley at any time".

Weissbach reported to Lerner that Genentech had the ability to do what the Institute could not. The question then became how best to structure a mutually-beneficial arrangement between the two companies. Recalling that conversation Weissbach says, "I said something like, 'Well, you can buy Genenetech.' He shook his head and told me, 'No, we're not going to buy that little company.' Then I said jokingly, 'Well, if you could get a scientist like Dave Goeddel...' and he said, 'No, we're not going to steal their scientist.' Really the only thing we could do was work out a licensing agreement with them."

While Herb Weissbach was very enthusiastic about working with Genentech, there was considerably less enthusiasm at corporate headquarters in Basel, Switzerland. The corporate Head of Research didn't share the American excitement about Interferon and wondered if this deal was necessary. John Burns and Ron Kuntzman got involved and, as Kuntzman said, "John just did what John always did, which was to smooth the ground and provide the support that was needed."

There also were some people in Nutley who were not in favor of this deal, as George Gould, who had to work out the details remembers, "We weren't quite sure how we were going to work together. My job was to work out some sort of letter of intent by which we would fund their research at Genentech and get the benefit of anything they discovered. There were some people inside the Institute who didn't want us to spend the money to help Genentech essentially get up and running. Sid Pestka, for example, was against it, believing there was nothing that Genentech could add to the mix that we couldn't accomplish at the Institute. He was adamant that we had the capability of cloning and expressing the gene without outside expertise."

While Udenfriend and Herb Weissbach respected Pestka's capabilities they knew that Genentech was far ahead in this specific technology. They believed the prudent step was to collaborate with a company that was already capable of doing the genetic engineering necessary to produce the high yield of cells that would produce large quantities of interferon.

Gould worked out an arrangement in which Roche agreed to pay Genentech several hundred thousand dollars a month to support their research. But after that deal was signed Pestka discovered that there were different types of interferon and the already complicated deal had to be renegotiated. Irwin Lerner wanted to include as part of the deal an option for Roche to purchase a stake in Genentech, but that company rejected it. So initially Roche and Genentech became partners in the development of two different types of Interferon.

Herb Weissbach was made head of a liaison committee between Roche and Genentech to develop interferon. John thought he was the best person to also smooth the relationship between Pestka's lab at the Institute and Dave Goeddel's people at Genentech. "I sent Pestka's clone out to Genentech," he says. "And within a month I got a call from Dave Goeddel on a Friday afternoon that they had it. They had modified the gene just as they had done for HGH and insulin, and they now had a bacteria that could make human interferon.

"I was pleasantly surprised how quickly they had done it. I sent Dick Stevenson, from John Burns office, out there to pick it up. He got back on Sunday; I took the sample home with me to make sure it didn't get lost. I put it in our refrigerator and told my wife, 'Renee, please do not throw this out.' On Monday morning I gave it to Pestka for testing. Their technique worked, the bacteria was making human alpha interferon.

This was the first interferon to be expressed in a mature, or complete form. But it still had to be purified from the bacterial extract. Theo Staehelin, from the Pharma Research Division at the Roche Institute in Basel, was working in Pestka's laboratory to produce monoclonal antibodies to interferon. He and Pestka worked through the weekend and successfully purified the interferon about 2 a.m. Monday morning. To celebrate this historic achievement Pestka prepared souvenir vials of the material for each of the people who had worked with him. Staehelin carried one of them back to his own lab in Basel and presented it to Roche International Chairman Fritz Gerber. When Gerber had it embedded in plastic to be kept in the company's historical archives they discovered it had leaked – and the sample was gone. It was quickly replaced.

There still were numerous hurdles to be overcome. "There were a lot of difficulties in the collaboration," Goeddel admits. "Some of them were Roche's fault and some were our fault because we were young and arrogant. We weren't the easiest guys to work with either. While Herb Weissbach wasn't doing the science, he was doing the dirty work and when there was a problem he was always there to help get it solved."

Both Pestka and Goeddel were brilliant scientists and they did not always get along. At times Weissbach had to be a mediator. "I could have been tough when I was young," Goeddel adds. "And while I will say that Sid Pestka was smart... I just didn't like his approach. He would give me selective data, when I thought that since we were collaborating we should both know everything".

What most surprised Goeddel was the respect with which he was treated by Burns and Lerner. "I felt I got even more support than I deserved from Roche management. More than I've ever seen in collaborations with other companies. When there was a conflict about anything that was scientific essentially they gave me the final say. I wasn't even 30 years old, so for me, at that age, that was pretty amazing."

And there were significant scientific challenges. The research proved to be quite different than they initially had anticipated. "Among the biggest surprises was that we expected the interferon gene to be expressed in very low amounts, and we'd have to search through a zillion clones to find it. But it turned out to be far less difficult than everyone thought it was going to be. Because we thought it was going to be difficult we used methods that were used to find a needle in a haystack rather than something that might be one of a hundred. In retrospect we agreed we should have been able to get this easier, but that often happens in science."

The result was a tremendously successful scientific breakthrough. According to Goeddel, "We thought we were good... we knew this technology. But Roche had built up a lot of expertise at the Institute in Nutley too. The combination was pretty powerful." In 1981 Pestka, Goeddel and Herb Weissbach published together; using Weissbach's in vitro protein synthesis system they demonstrated that the plasmid carrying the human interferon gene could synthesize the human interferon protein in a test tube.

While the Institute and Genentech were working together, they still were in competition, and there was considerable debate about who was responsible for the development of Interferon. Obviously Roche scientists claimed credit, as did people at Genentech. In fact, on June 4, 1980 Genentech issued a press release announcing that it "had successfully produced the two types of human interferon, fibroblast and leukocyte, through recombinant DNA technology. These interferons have been reported as potential antitumor and antiviral agents. Collaboration on this project was undertaken with Hoffmann-La Roche.

"Scientists of both companies have been working together for over a year in pursuit of large scale commercial production of pure interferon by means of recombinant DNA technique. This was rapid progress since in 1978, less than 3 years ago, scientists at the Roche Institute of Molecular Biology were the first to succeed in producing pure interferon from human leukocytes in amounts sufficient for physical and chemical analysis."

At the same time Pestka reported the success in RIMB's 1981 annual report, writing, "Because of our success in purification, determination of the interferon structure and in cloning and expressing human interferon in bacteria, clinical trials have begun with pure interferon obtained by recombinant DNA technology."

While it clearly was a joint effort, Weissbach remembers "Sid Pestka really felt he had found this clone. He isolated it and it was his, and that credit was being taken away from him. He believed strongly that he could've done the entire process in his lab if he had been given a little more time." Pestka's role was recognized in June, 1985, when the New York Times reported, "Hoffmann-La Roche Inc., the subsidiary of F. Hoffmann-La Roche & Company Ltd. of Basel, Switzerland, said it had received the first United States patent for alpha interferon. There are three classes of interferon. Hoffmann-La Roche said it had patents on all 12 types of alpha interferon, which is produced by white blood cells." Pestka and Menachem Rubinstein were each paid the usual $1 for their work.

Obtaining that patent had been extremely complicated, as patent law for biotechnology products was still evolving. "It took us seven years," George Gould said, "but eventually we were able to convince the patent office that no one could predict such activity from anything that was done previously until it had been purified and separated, and then it was possible to measure the effect of interferon on different cell lines." The work done inside the Institute allowed these strains of interferon to be protected by patent.

The fight was well worth it. Gould estimates that "Interferon, in all its forms based on the natural patent that came entirely out of the Institute and then the Genentech recombinant studies, which was the result of the collaboration between Pestka and Goeddel, brought in eight or ten billion dollars in revenue, maybe more."

While interferon has been an extremely profitable product, its actual benefits have proved to be somewhat inconclusive. For a time it was considered the new miracle drug and many people believed it had the potential to be the cure for cancer, and while it does have some positive benefits in the treatment of some cancers, it has yet to fulfill the early promise. There was considerable disappointment throughout the medical community when it did not fulfill the early optimistic promises, but it has been shown to be an effective treatment in conjunction with other drugs in diseases like multiple sclerosis, certain types of cancer and other viral diseases. It also has provided a window into the creation of new therapies, and research on its potential is still being done.

But this project proved far more important than simply being a financial success. Interferon-alpha became the first Roche corporate product created from the new recombinant DNA technology. It also served as a prototype for the many other biotech products that would follow. The existence of RIMB made it possible for Roche to become one of the very first – if not the first – major pharmaceutical corporation to take the giant step into the emerging world of biotechnology. This collaboration with Genentech on interferon proved to be only the beginning of what went on to become the most significant corporate cooperation in the almost century-long history of Hoffman-La Roche. While the two companies collaborated on this success, as well as on several other projects, there also were some rocky moments. In other areas, like growth hormones, they competed with each other for similar products and the same clients. That competition became a legal fight in 1986 when Hoffman-La Roche sued Genentech for "conducting clinical trials on synthetic human growth hormone formulated in pharmaceutical preparations designated 'Protropin' and submitting (an NDA for the product) to obtain approval . . . "

At the same time Genentech had hired several Roche corporate and Institute scientists. Among them was a research assistant who had worked in Pestka's lab and several post-docs. But additional problems occurred when Genentech wanted to scale-up its manufacturing capability and hired two key Roche corporate employees from the Roche Biopolymer facility, located on the 5th floor of the Institute building. The Roche employees that Genentech hired had expertise in scaling up fermenters to grow large amounts of bacteria and purifying the recombinant proteins. That caused Roche lawyers to threaten legal action. As a result, Dave Goeddel admits, "Swanson came into the lab and said, 'Okay, we'd better not hire any more Roche guys. I just got threatened.' And that was the end of that problem.

While the euphoria at the birth of the biotech revolution cooled as the promises diminished, the hard work of building a real foundation to support the new scientific technology was being laid. Although several of the once well-funded biotech start-ups were unable to compete and either merged with others or closed down, Genentech emerged as one of the leaders of the industry. By 1990 it had created four of the first six biotechnology drugs to have received FDA approval. "There was no question our work had earned a tremendous amount of success," Goeddel says. "Early on, when we did a few things, several companies started on the premise they could do it too. But by the mid-1980s we'd established our track record and I heard from a lot of people at a lot of companies that they had strategy meetings to figure out what we weren't working on so they could go after it."

But even with its enormous success Genentech was struggling to cover the substantial costs of research and development. While its revenue stream from its products was growing, it wasn't sufficient to cover day-to-day expenses. So they began scouting around the world to find a strategic partner. "Herb Boyer was looking for money," Roche US President Irwin Lerner recalls. "He was visiting a lot of big companies in Europe, one of which was Roche, of course. The then new chairman of Roche, Fritz Gerber had come to Roche from the insurance industry and didn't know a lot about the pharmaceutical business. After meeting with Boyer, Fritz Gerber told me he was very impressed and he wanted to buy the company.

"'Wait just a second,' I said to him. 'Why do you want to spend hundreds of millions of dollars to buy the whole company when you can buy a piece of it and, as a practical matter, own it anyway?' By buying only a piece of it we were putting ourselves in a strong position. If anyone else dared to come near them to make a takeover offer we could move to buy the remainder of the company."

All this was taking place while the image of the entire industry was changing drastically in the public's mind. The involvement of hedge funds and venture capitalists had shifted the focus of the industry from scientific achievements to bottom line concerns. While once the industry had been viewed quite positively by the general public, pharmaceutical companies more and more were being viewed as cold and avaricious companies. "George Merck, who founded Merck and Company, had a motto," Lerner explained, "Medicine is for the people, not profits. But when the money people got involved all that mattered were profits; suddenly people in the industry were talking about earnings per share, bottom line results. While I don't think that was fair, I do think the industry's attitude changed from searching for new products to generating as much as possible for the bottom line. That led to a lot of practices that brought the industry so much criticism. I recall saying at one point 'Thank God for the oil and cigarette industries or we'd be the number one target."

Those public companies with their focus on increasing stock value in an extremely competitive industry had a far more difficult time making an acceptable offer to Genentech than Roche. For the same reason that Roche was able to decide so quickly to create and support the Institute, this was a family-controlled company, they were able to make an offer to Genentech.

Throughout the 1980's Roche had been buying and selling and buying Genentech stock, but more for investment purposes than to form any type of permanent alliance. But finally the decision was made to take a substantial stake in Genentech. It was, at the time, the single largest investment that Roche had made in another company in its history. In September, 1990, "Genentech

Inc. and Roche Holding Ltd. Today agreed to a pioneering agreement in which Roche will invest approximately $492 million in capital into Genentech and purchase half its stock for $36 per share." The entire deal cost Roche $2.1 billion for which they received 60% of the equity in Genentech. Roche also obtained an option to purchase all the outstanding shares.

Announcing the deal, Genentech co-founder Robert Swanson said, "Genentech will have the resources and the independence to fully deliver the dream of biotechnology. I truly believe that by the turn of the century essentially every new drug will be touched in its development by biotechnology"

There were many people at Roche who had doubts about the wisdom of the purchase. Steve Krognes, then a Roche executive, told the Wall Street Journal, "It was a highly controversial move in Switzerland. A lot of people thought it was a waste of money (to buy) a young biotechnology company for $2 billion. But there were a couple of people at the top of the company (who said)... 'This is the future of (Roche).'"

There also was quite a bit of trepidation inside Genentech, as many employees there believed that along with stock they had sold their independence and creative freedom. The very loose entrepreneurial Genentech culture created by Boyer and Swanson was quite different than the mostly serious way of conducting business at Roche. Ronald Martell, a former Roche employee who went to work at Genentech in 1988 recalled that as he walked into the door the first person he met was a man wearing a grass skirt and a coconut bra telling people, 'Surf's up!' He took a deep breath and dismissed it, knowing full well that every company has its characters – until he discovered that this person happened to be Bob Swanson. Swanson was simply in costume for the regular Friday afternoon company beer fest.

Some Genentech employees took pride in the fact that they had taken off their tie their first day there and never put it on again. Now they were concerned that this spirit might be dulled. To try to calm those fears Fritz Gerber issued a statement, saying "Roche's $2.1 billion total investment will be best served if Genentech continues to operate in the manner that led to its impressive success in such a short time. We are committed that Genentech proceed with its business and maintain its enterprising spirit with autonomy. Both companies will retain independent control of their research agendas and continue their existing business relationship with other pharmaceutical and biotechnology firms." He did not say a word, however, about grass skirts and coconut bras.

It is probable that at least some of the fears of Genentech scientists were calmed by the existence and the history of the Institute. Roche had guaranteed complete intellectual freedom to the Institute when it was founded and had followed through admirably. There was never a time that the corporation had attempted to impose its will – or its needs – on the research carried out in the Institute. It had been, and continued to be, a very productive relationship.

Inside the Institute there also was a combination of delight and concern. Without question it had been the Institute's initial relationship with Genentech in the co-development of interferon that had brought the companies together and shown the potential strength of the combined entity. The purchase could be directly attributed to the existence of the Institute; conversely, there was some trepidation inside the Institute about the areas of overlapping research between Nutley and Genentech.

There was no way of knowing what form this new relationship with Genentech would take and what its' impact would be on the Institute. The world of biotechnology had changed drastically in the two decades since the founding of the Institute. In the late 1960's the concept of a center devoted purely to research was new to the pharmaceutical industry and many people doubted it would survive. Since then an entire new segment of the industry had been created, the biotech start-ups, which had successfully bridged the wide gap between academia and industry. So Roche's continuing need for the Institute was less obvious.

The fact that several of Roche's most successful products, as well as the initial relationship with Genentech, were due to the Institute might too easily be overlooked. Irwin Lerner liked to tell a story about his relationship with the great scientist Leo Sternbach, whose discoveries led to the creation of librium and valium, as well as chemically synthesized vitamin E, and vaulted Roche from a mid-sized manufacturer of vitamins to a pharmaceutical giant. "The guy was a genius," Lerner said, "and I was smart enough to stay glued to him. I used to bump into him in the elevator in the research tower and every time I would see him I would say, 'So Leo, what the hell have you done for Roche lately?'" And that was the question that most concerned people inside the Institute, after the purchase of Genentech, were worried they would be asked, "what have you done for us lately".

By the time this deal was made there already had been substantial changes inside the Institute's own interferon team. Menachem Rubinstein, who was credited with developing the reverse phase high-performance liquid chromatography (RP-HPLC) method that Pestka had used to isolate interferon, and was cited as a co-inventor on the patent, had left the Institute and returned to Israel soon after the patent was filed.

And in 1986 Pestka also had left, accepting a position as Professor and Chairman of the Department of Molecular Genetics, Microbiology and Immunology at UMDNJ-Robert Wood Johnson Medical School in Piscataway, New Jersey, a State run research Institute.

Following the suggestion of the Board of Scientific Advisors, Herb Weissbach had begun encouraging turnover to make space available for younger scientists. In fact though, the only person Weissbach pressured to leave was Severo Ochoa, "and I did that because both he and his wife, Carmen, were physically going down hill. I had spoken to several of his former post docs in Spain and they really wanted Severo to come home. If he returned to Madrid, they told me, they would make certain the cost of all of his medical issues would be taken care of. Additionally, it was obvious that the daily commute from Manhattan was becoming more difficult for him. 'It'll be better for you back there,' I told him and he finally, and reluctantly, agreed. It was hard for all of us to see him leave; his love for science was still strong and he knew he would not have the facilities or support he needed to continue working in Spain."

No one else had to be pushed, by that time most of the senior scientists had achieved international recognition and were being actively recruited by universities. In addition to Pestka this included Aaron Shatkin, Ron Kaback and Syd Spector. Pestka's leaving was not the easiest departure; with the company's permission Pestka had taken all of his original notebooks out the door with him. This became a problem years later when Roche was involved in complex litigation and, as Gould says, "Without the original notebooks we were really hard pressed. I had to go back and talk to one of his lawyers who essentially blackmailed me to try to get access to the books."

Pestka had been one of the original members of the Institute and was responsible for numerous patents in addition to interferon. For his work, most of it done at the Institute, he eventually was inducted into the New Jersey Inventors Hall of Fame and in 2002 was honored by President George W. Bush at a White House ceremony at which he received the National Medal of Technology for "pioneering achievements that led to the development of the biotechnology industry, to the first recombinant interferons for the treatment of cancers, leukemia's, viral diseases such as hepatitis B and C, and multiple sclerosis; to fundamental technologies leading to other bio-therapeutics; and for basic scientific discoveries in chemistry, biochemistry, genetic engineering and molecular biology from protein biosynthesis to receptors and cell signaling."

Hoffmann-La Roche's relationship with Genentech, which was a direct result of Pestka's research on interferon, continued to grow after that initial purchase and eventually caused fundamental and permanent changes in the entire company. As was reported many years later by Roger Longman in the popular industry blog, In Vivo, "(T)his has been the most successful relationship in pharmaceutical history. Neither company would likely exist as an independent entity without it. Genentech was an acquisition waiting to happen back in 1990 when it managed to keep at least managerial independence by selling Roche 60% of its shares. Without Genentech, the Swiss giant would most likely be part of another Swiss giant, Novartis (which still owns a small stake in the pharma). It's worth remembering that in 2007, Roche got 28% of its sales from Genentech-sourced products... Moreover, one-third of its Phase II and Phase III pipeline is comprised of Genentech programs. Meanwhile, Roche's 56% ownership of Genentech accounts for roughly a third ($55.2 billion) of its $154 billion cap, the second biggest in Pharma after Johnson & Johnson.

"In case you aren't getting it, let's be clear: Roche's success is largely due to Genentech."

307

And without RIMB, this was a relationship that would not have existed.

In fact, contrary to the fears of the Genentech people Roche management never tried to impose its corporate culture on them. Perhaps because Roche had seen the benefits of an independent research Institute they did not make changes; for example, just like the Institute, Genentech encouraged its scientists to publish their discoveries in major science publications rather than keeping them secret, a policy that continued after the acquisition.

The relationship between the two companies went through several different stages after the initial purchase as management tried to find the right fit. In 1999, for example, Roche exercised its option to purchase 100% of Genentech – but a year later turned around and offered more than 17 million public shares of the biotech company for sale – thus earning back almost the entire cost of the acquisition while still retaining a 58% ownership stake in it. But from that deal Roche also retained the rights to license and market Genentech products around the world.

In some ways, as members of the Institute might have feared, Genentech began filling the somewhat similar research space. That might well have figured into the corporate decision to close the Institute in 1995. In the ensuing years Genentech produced many important biotech drugs, including Rituxan to treat forms of cancer and arthritis, as well as the active cancer drugs Avastin, Herceptin and Tarceva.

The financial relationship between Roche and Genentech changed several times after the turn of the 21st century, as Roche management bought and sold shares, until in 2008 when Roche announced its intention to buy all of Genentech. One reason for that decision, according to Steve Krognes, a former Roche executive who became the CFO of Genentech, was that "The Roche business in the U.S., frankly was not doing that well. Nutley has a long and lustrous history and a lot of great research has come from there. But Nutley has not produced anything for a long time. The United States represents 40% of the world market. If you have ambitions to be a leader in this country, not to have a good presence in the U.S. is not acceptable." That 'long time' certainly doesn't mean only the decision to close the Institute, but there is no question that doing so damaged Roche's research capability and it became more and more reliant on Genentech-discovered products.

The deal was finally concluded in March 2009, when Roche agreed to pay $46.8 billion to complete its ownership of Genentech. It is fair to say that while valium is the drug that built Roche into a pharmaceutical giant, one of the great legacies of the Roche

Institute of Molecular Biology was the relationship that was formed with Genentech.

Chapter Seven

The Supporting Cast

While the actual structure of the Institute tended to be somewhat informal, there was a typical hierarchy. The senior scientists, such as Udenfriend, the Weissbachs, Shatkin, Pestka and Spector and others at that level were referred to as Full Members, which in an academic setting was the equivalent of a Full Professor. They were permanent employees of Roche; the closest equivalent of having tenure. Associate Members also had tenure, but the less experienced independent investigators, who did not quite have the same level of experience or achievement were Assistant Members and they were given five year contracts. They would be evaluated after 3 years and in many cases were promoted to Associate Member within 5 years. There usually were several visiting scientists, although Severo Ochoa was the only Distinguished Visiting Scientist. The largest block of people were the postdoctoral fellows and there generally were several very bright young predoctoral fellows, graduate students conducting the research required to complete their thesis. At times honor students from colleges or even, in a few cases, high schools would be invited to work at the Institute. In addition, there was a complete and experienced support staff which included animal services and glassware services.

Because most people enjoyed working at the Institute turnover of the senior staff was small during the 1970s. By 1983, when Herb Weissbach took over as Director, the Board of Scientific Directors expressed concerns to the Institute Executive Committee and Roche management that the Institute could become somewhat stagnant if there was not turnover of the "tenured staff." This was a delicate situation, but that message to the senior staff encouraging them to explore exciting new opportunities, which many were receiving, did get through. By 1986 more than 50% of the original members of the Institute in 1971 had left, and by the late 1980s the average age of the senior staff was 43, a decrease of 20% from its high.

Initially, there were three large departments; Biochemistry, Virology and Physiological Chemistry. These areas of research in the newly formed Institute reflected what the Executive Committee felt was right in 1967. But as new discoveries in molecular biology and molecular genetics pushed the emerging biotech industry into different and sometimes unexpected areas, the structure of the Institute was changed to reflect that and encompass the growing diversity. And at least some of the areas being investigated did not fit easily into any departmental structure. Frank Margolis, one of the first scientists Udenfriend recruited, was working on the biochemistry of smell, specifically the difference between olfactory cells and other brain cells. Why and how, he wondered, can you instantly recall an early childhood smell? By 1980, for example, the Department of Physiological Chemistry had become Physiological Chemistry and Pharmacology.

By 1991 the emphasis of research had changed drastically and the Institute had been restructured, on paper at least, into four departments that demonstrated this: the original Department of Biochemistry was gone, illustrating how important specialization had become. It was now called the Department of Gene Regulation – which was formed to research signal transduction, gene expression and gene regulation, The original Department of Physiological Chemistry was now the Department of Neuroscience, indicative of how important the Institute felt this field was, the original Department of Virology was now the Department of Cell and Developmental Biology, and a 4th Department was now present, called, Molecular Oncology and Virology.

Finally, by 1993 it was obvious that similar technologies had expanded into so many areas of science and research that departmental names simply no longer made sense, and they were eliminated completely. As Director Herb Weissbach explained, "It has become clear there is so much common interest and interaction among scientists in the various areas that for practical purposes the Institute has been evolving towards a more integrated structure... (Recognizing that) in 1993 we eliminated the departmental structure. Since the inception of the Institute we have had departments with names that reflected the interests of the scientists housed in that department. However, with the research areas changing so rapidly over the years these names were losing their significance. Also, without a departmental structure it will be possible to make better use of space and to more easily relocate groups to increase scientific interactions..."

While the essential purpose of the Roche Institute was to become a place where leading scientists in the world could pursue their curiosity, it also was established as a place for those scientists to train the next generations of researchers. As the charter stated, 'The Institute will also provide an opportunity for promising young scientists to obtain advanced training and to engage in fundamental research under conditions encouraging professional growth and creative effort."

Most postdoctoral fellows are more concerned about the scientist in whose lab they will be working rather than the institution. Training under a well-respected scientist is an important step in launching a career, so Members of the Roche Institute, with few exceptions, received more applications then they could handle. The fact that the Institute also paid considerably better salaries than most other laboratories also boosted recruiting, and as a result the postdocs tended to be really smart, creative and committed to science. The postdocs were the life blood of the Institute, eager to move their career forward, always on the move, and at times there seemed to be an endless supply of them. They were young scientists who had finished their Ph.D's, and quite often had done research for their doctoral degree in the same field or adjacent fields as their mentor at the Institute. RIMB was the launching pad of their careers. Silvio Spadari said simply, "(Art Weissbach) was able to turn an enthusiastic apprentice into a dedicated professional and the two years I spent in his laboratory definitely represented by far the most important event of my scientific life... Those years have influenced my entire career more than any other circumstance allowing me to shortly climb all the steps of my career."

"The contributions the Institute made to its employees," said Art Bolden, "and in particular to personal growth of people like myself, looms large in my domain and stands out like a beacon in the forest."

A postdoctoral fellowship is a key period in the training of a young scientist. Essentially, it is a period of several years during which they work in a laboratory, under the tutelage of a proven scientist, both picking up knowledge of the field in which they intend to work, but equally important learning the techniques of conducting productive research. According to Nik Robakis, a former Institute postdoctoral fellow with Herb Weissbach, who now holds an endowed chair as a Professor at the Mt. Sinai School of Medicine, where he is doing cutting edge Alzheimer's research, "Science is all about asking the right question at the right time, and having the tools and the right environment to be able to answer that question.

"A scientist has to learn how to recognize and understand the process. The mind of an observer has to be trained so that when the moment of revelation occurs, he or she is going to be able to see it and interpret it."

Postdocs also, Robakis adds, provide necessary cheap labor for an institution. "They benefit by learning, and society benefits by having young, curious, energetic scientists working very inexpensively. Postdocs are the worker bees. The fact is that without them it would not be possible to make the progress that has been made in biological science."

There was a postdoctoral Committee and George Cardinale, a former scientist turned administrator, was hired to help the postdocs in any way possible. Because of the culture of the Institute the postdoctoral fellows felt comfortable getting expert advice from people in other laboratories, so there was a constant, uninhibited flow of information among the different laboratories.

That 'training mission,' was an essential aspect of the Institute – and the fact that it existed at all was a tribute to Roche President Barney Mattia. The concept that young people would work at the Institute for two or 3 years and then would usually leave never made sense to him. Eventually though, he agreed to the concept that training should be one of the missions of the Institute.

No one embraced that training aspect more than Udenfriend. As Roche Research Leader Arthur M. Felix wrote, "Sid would often bring his post-docs to meet with my colleagues and me... He imparted so many of his scientific qualities to his students (predoctoral and postdoctoral) and to so many colleagues, that he is part of many of us."

As Udenfriend wrote proudly, "To have an impact on the world of science, the staff of the institution must do more than carry out research. The Roche Institute of Molecular Biology is, in fact, a miniature academic center. One of the major roles of an academic center is the training of young people. Most of this is done through the Postdoctoral Fellowship Program which has a worldwide reputation. What is not known so well is our Predoctoral Training Program, which is small but high in quality. The Institute functions as a regional graduate center with students from several universities in the New York/New Jersey metropolitan area. The first year is generally devoted full time to course work at the university. By the second year students are involved in full time research at the Institute. The member of the Institute who is the student's mentor holds an appointment on the faculty of the appropriate university..."

"From the initial discussions in 1967," wrote Bernard Witkop and Herb Weissbach in a biographical memoir for Udenfriend for the National Academy of Sciences, "it was clear that the long term success of the RIMB as a basic research center would depend on being able to attract postdoctoral fellows and graduate students. Sid Udenfriend was determined that this would be the case... At that time universities were reluctant to accept industry scientists as adjunct faculty and there was a period of great concern...that Institute scientists would not have university affiliations and thus not be able to have graduate students... Eventually RIMB scientists had appointments at most of the large

318

universities in the New York-New Jersey region. Postdoctoral fellows were anxious to come, and there was no aspect of the RIMB that Udenfriend was more proud of..." During the existence of the Roche Institute almost 1,000 fellows and more than 40 graduate students were trained there. By the time the Institute was closing in 1995, 20 of these trainees had become Chairmen of university academic departments, 50 were full-time Professors and 60 had become Associate Professors. One man had become a university Vice-President, and 35 were in leadership roles in the pharmaceutical industry.

As the founders had envisioned, as much as the Institute would do for the young scientists, their presence there also was vital to the success of the Institute, "These Fellows bring to the Institute new ideas, highly imaginative approaches to problems and a degree of enthusiasm and motivation for seeking answers, which provide a constant source of scientific stimulation and drive for the permanent members of the staff. This continuous influx of Fellows, who perhaps are among the very best scientifically, forms a base from which the most profound observations and achievements can be realized for the future."

The number of postdocs working at the Institute at any time was flexible. Each investigator had about five or six people available to them, which meant there were about 30 on each floor.

319

The primary qualification was a good academic background and a strong desire to learn. "We were considered temporary employees," Joe Tarnowski explains. "But they gave you nominally a two-year commitment that could extend into three, but their goal was to train you, expand you, get you a few publications and push you to move on with your career. They even paid reasonably well compared to academic post-docs, the salary was as much as 25% or 30% better."

Barbara Kerr, who worked at the Institute for 18 years as a secretary, assistant and whatever else was needed, remembers that the new postdocs would arrive each year in early July and for the next few weeks she would constantly be running into people wandering through the halls, completely lost, trying to find their way around the place.

Among the traditions for new postdocs was what Ron Kaback described as his "advanced interviewing technique for new postdocs: After they were in the lab for a week or so I would invite them to my office for a conversation. Under my desk I had hidden a fire extinguisher. The new postdoc and I would be in the middle of a relatively casual conversation when I would reach down and squeeze the release mechanism – which would blow out carbon dioxide under the desk in a great plume of smoke accompanied by a big, loud bang! The postdoc would hit the ceiling." As the postdocs settled in they would on occasion find ways to get their revenge.

These postdocs came from around the world to work and study at the Institute. There was simply no way of making any generalizations about them – except they were young, enthusiastic and smart. Some were married, many more were single; some were financially set while others were struggling. Their interests were extraordinarily varied; Karl-Werner Knopf from Germany deciphered hieroglyphics in his spare time. Barry Garfinkle collected stamps and built scale models of sailing ships. Rich Adler was a civil war buff and spent his off-time exploring battlefields.

Nat Brot, for example, had a graduate student from China working in his lab. "Her parents had both been academics who were caught up in the Cultural Revolution," he remembered. "Her parents had been sent out to work in the fields and she had been left home alone as a 10- or 11-year old. She would tell me how she had to carry bags of groceries up several flights of stairs to the apartment. Her parents contacted Herb Weissbach asking for an opportunity for their daughter to learn at the Institute. Herb didn't have a position open for her, so he asked me if I would take her both in the lab at the Institute as well as at the University of Medicine and Dentistry of New Jersey where she was enrolled as a graduate student. She earned her Ph.D, and then went on to work in industry. But she was typical of the young, smart, enthusiastic people we would have come in."

Tim Coleman had an unusual experience; he came to the Institute right out of college. "We were technicians. I'm proud of it. I would never change it for the world. I learned from the bottom up. I did mundane jobs, I did the jobs nobody wanted. I didn't care, they were paying me and I was getting an opportunity to work. We were there to support the real scientific inquiry. I worked in Herb Weissbach's lab, and he would tell me, 'Tim, I want you on this project working with (so-and-so). Here's what we're trying to do, so have at it.' The routine jobs were part of it, I had to maintain the centrifuges and the rotors. I had to make sure there was liquid nitrogen and maintain the -80 freezers. As new postdocs came in I had to show them where to find things.

"And then I was expected to do hands-on research. I was expected on some routine basis to stand up in a group meeting and explain what experiments I had done, what results I got and what I thought I should be doing next. Then we would talk about it. And as I got better and as our projects moved along I was expected to create posters to present at meetings, and write papers and even make presentations at larger meetings. But every step of the way I had every opportunity to learn from all those brilliant people. There were stars in that building. For many of us, that was where we learned about science. Like several other technicians I worked with we all ultimately went back to school to get our Ph.D's. I can't tell you how many of us became so inspired by the science that we learned there that we had to do more."

Tim Coleman did just that and while in graduate school at Ohio University remembers that he had several offers coming out of graduate school, "I had offers from Johnson & Johnson, from Sandoz that paid better, but Herb Weissbach advised me, 'Don't win the battle and lose the war. Come here, you'll win the war, and so I went back to the Institute".

The career path of John Kopchick was typical of so many of the young scientists who learned their craft in Nutley. Kopchick had been studying molecular retrovirology, the viruses that cause cancer, at the famed MD Anderson Hospital in Houston, Texas. "When I told my advisor at Anderson I was considering doing postdoc work at the Roche Institute he was ecstatic. Initially I hadn't known much about it. I responded to an advertisement looking for postdocs and went up to New Jersey in early 1980 for an interview. I spent two days there meeting with the senior faculty. I thought I was pretty good, but when I got there I met these people who knew molecular biology; they ate it, they breathed it, they lived it 24 hours a day and I knew this this was the place I wanted to study. When I went there you couldn't

Google people's names to learn how successful they were, but as I reviewed the literature and names like Aaron Shatkin, Sid Pestka, Sid Udenfriend, Herb and Art Weissbach, you would see that these were stellar individuals in terms of scientific publications, industry recognition, invitations to lecture worldwide.

"In addition to the people working there, the facility itself was outrageously wonderful. It was a multi-floor building, which nobody had ever heard of, on top of a hill with every piece of equipment imaginable." Working there, as a postdoc, fulfilled his expectations. "Probably more than anything else, most of the people at the Roche Institute had a passion for learning. It didn't matter who they were, or what they had accomplished, they had this hunger to learn more. We would have the opportunity to meet with and work with these world class scientists, not monthly, not weekly, but every day. For example, my hero was Aaron Shatkin who was as good as anyone in the world in terms of molecular biology. And we would, we, meaning he and several others on the senior faculty and us postdocs, would sit around and have a bagged lunch every day. We'd all bring our lunch and sit there and talk sports, politics and science, a lot of science. It was very informal and we got to be good friends with some of these people. It was much more than a boss-underling situation, we were treated as colleagues. And every once in a while we actually would be able to make a contribution. We would make a suggestion or ask a question that would make them think in a new direction."

As Kopchick emphasizes, postdocs were welcomed to the Institute with a reminder of its objective, "None of the faculty ever told us, hey we've got to discover this so that we can have a new therapeutic target, never. It was all basic research that hopefully would lead to publications. To spread the knowledge so other people could benefit from it. That was the mandate we postdocs were given. Now if something happened to be so exciting that it could be considered a therapeutic target or perhaps a therapeutic in itself, like interferon, that was great, that was icing on the cake. But they reminded us that was never the objective of the Institute.

"I just will never forget that feeling of excitement when you walked into the building. Every day. That doesn't mean that every experiment was successful, every person there would have their scientific-slash-academic downtime experiments that didn't turn out the way they wanted, every scientist goes through that. But there were enough people that had good results that there would always be excited people running around the Institute. And when you walked by a lab you could hear easily the excitement, 'Hey great!' and people giving each other high fives. Or you could hear 'Oh dear' in the hallway. Most importantly, you could go into any lab and ask a postdoc or a technician or senior leader of the department, whether it was in my case Art Weissbach, or Ann Skalka or Pat Gage or Aaron Shatkin, 'I don't understand this method of cloning, could you teach me?' and they would say, all of them, 'Yes, come in, come here for two weeks and just learn how to do it.'

"There weren't little kingdoms ruled over by a benevolent leader. The senior scientists wanted the postdocs to learn not only from them but also from other faculty engaged in the types of experiments that we were doing. These large groups interacted very well with each other.

The opportunities to learn were available to any young scientist willing to seek them out. According to Joe Tarnowski, "I worked for Herb Weissbach and his postdoc cadre in his lab. We all worked together but there was the expectation that if you needed expertise, you go down the hall and there's another principle investigator who was a specialist in that area who would help you. The Institute was set up to be horizontally and vertically integrated to allow as much cross-fertilization as possible. You could just go up and down the hallways and stair wells and find an expert on anything you needed to know about molecular biology and protein chemistry."

As defined in the charter, postdocs weren't simply seen as additional hands to do the grunt work in the labs, they were treated as collaborators in the exploration of the science. They were there to learn how to be active, productive scientists; not only how to conduct an experiment but how to become part of the greater scientific community. Just as with the senior staff members, there were no tests, they didn't have to fulfill any requirements and rarely was anyone asked to leave. But they were encouraged to actively participate in the work of the Institute. They were encouraged to write abstracts; several paragraphs outlining the work they were doing, which were sent to organizers of meetings around the country – and almost invariably they were invited to participate in those meetings. Kopchick, for example, had become proficient in micro-injection techniques, which he wrote about. His paper was accepted by people organizing a conference in Aspen, Colorado. He remembers getting into a car with his supervisor, Dennis Stacy, and driving from Nutley to Aspen to help conduct a one-week course for 20 scientists from around the world who had come there to learn this technique.

After working and learning at the Institute for 2-3 years the postdocs were encouraged to take their knowledge and experience into the worlds of academic and industrial science, sometimes both, and their achievements have been extraordinary. The list of young scientists who trained at RIMB then went on to make substantial contributions to research and scientific education is seemingly endless. Picking a representative sample of those people is simple and serves to illustrate the lasting effect of the Institute. Not only did many of these postdocs do important and meaningful research, they also passed the baton, training others to the standards they learned at Nutley.

After completing his studies at RIMB, for example, John Kopchick turned down an offer to work at Roche and accepted a position at the Cleveland Clinic – where he worked in Amiya Banerjee's department. He later went into academia, becoming a Professor of Molecular Biology in the Department of Biomedical Sciences at the Ohio University Heritage College of Osteopathic Medicine. His ground-breaking work in the field of growth hormones impacted the study of endocrinology and he discovered the drug Somavert which is used to treat a disease called acromegaly which is characterized by the excessive secretion of growth hormones.

Numerous postdocs have taken the training they received at the Roche Institute into both academia and industry. The list of young postdocs whose distinguished careers were born at the Institute is long and impressive, and it would be impossible to cite them all. But among a sampling of the hundreds of men and women who laid the foundation for their career at the Roche Institute is Diane Pennica, whose work with Herbert Ennis on RNA transcription and translation attracted attention at Genentech and after completing her postdoc work accepted a position there. She began cloning proteins, eventually becoming a Senior Research Scientist. She collaborated with David Goeddel and several other scientists on the creation of a very important heart attack drug, and eventually being honored with three other people as the 1989 Inventor of the Year by the Intellectual Property Owners Foundation. In her career at Genentech she has published almost 100 papers and received 20 US patents.

One of the first postdocs to join the Institute was Amiya Banerjee, who had earned his Ph.D. at the University of Calcutta and came to the United States to finish his studies in molecular biology at the Albert Einstein College of Medicine. "Molecular biology was in its embryonic state and I was studying bacteriophage, viruses that infect bacteria rather than human cells," he remembers. "We made some very good observations there and when I was getting ready to go back to India none of the laboratories there were using bacteriophage. So I decided I would learn something more about animal viruses, then go back. In 1968 an advertisement on our bulletin board that the newly started Roche Institute was asking for post-doctoral fellows to come learn, and work with the outstanding faculty caught my eye. I decided I should go to this Institute and learn animal biology. I looked at the faculty and it was Arthur Weissbach working on herpes and other DNA viruses, Aaron Shatkin working on retroviruses, Brian McAuslan working on vaccinia and I thought, this is a great department to join.

"I submitted my application and I soon received a letter from Aaron Shatkin, who wrote he was very interested in me, and maybe I can help him set up his laboratory. My friends in academia who were always struggling to get the right grant thought the whole idea of this Institute was totally bizarre. But when I went to work there it was nirvana; we knew we had to work extremely hard because Sidney Udenfriend had created a big philosophy, a golden opportunity to work hard and contribute to science.

"I was supposed to stay for only a few years, but when my time was completed they invited me to stay. After a few years RIMB gave me my own laboratory. Aaron Shatkin was my mentor but became my colleague. I was working with him on the discovery for which he became a member of the National Academy of Science, and when he was honored that was one of the greatest days of my life too. I stayed there for 18 years until I was offered the opportunity to become Chairman of the Department of

Molecular Biology at Lerner Research Institute at the Cleveland Clinic".

In his career Banerjee has had almost 250 articles published, and his work has been cited in other research almost 9000 times.

Former postdoc Joe Tarnowski holds 13 patents with several more pending; after leaving Roche he worked at Bristol Myers Squibb as the Senior Vice President Biologics Manufacturing and Process Development, where he supervised construction of that company's $750 million large-scale multi-product bulk biologics manufacturing facility in Devens, Massachusetts, and the integration of Medarex, Inc. He then became the Senior Vice President of Chemistry, Manufacturing and Controls (CMC) Biopharm R&D at GlaxoSmithKline.

Grace Ju accepted a position at Roche after working in Anna Skalka's lab, and through the years directed preclinical research groups in the departments of Molecular Biology, Immunology, Oncology and Inflammation, eventually becoming the Director of Inflammation Discovery research. She has authored more than 45 publications and has been an invited speaker to numerous conferences and symposiums.

In some ways Lynn Enquist completed the original circle when he accepted a position at the NIH after leaving RIMB, going to the government Institute that had served as the incubator for the founders of the Roche Institute. In 1993, after a career as a Senior Research Fellow working on developing neurotropic viruses as tools for gene therapy and studying the mammalian nervous system, he accepted the position of Full Professor of Molecular Biology at Princeton University and eventually became Chairman of that Department. He has become one of the nation's leading virologists; he is the former President of the American Society for Virology and is the author of more than 275 articles and books, as well as holding four patents.

Brian Cullen came to the Roche Institute from England, and worked there while obtaining his doctorate from the University of Medicine and Dentistry of NJ, and eventually was honored as the James B. Duke Professor and Director of the Center for Virology in the Duke Department of Molecular Genetics and Microbiology. Among his numerous contributions have been the discovery of the mechanisms of action of the essential HIV-1 proteins Tat and Rev and his demonstration of the key role played by small non-coding RNAs in regulating herpes virus latency.

After working with Tom Curran at the Institute, Cory Abate-Shen became a leader in the fight against urological cancer. Abate-

Shen created the Prostate Program for the Cancer Institute of New Jersey and served as its co-leader from 1999 to 2007, then accepted a position as director of research in the Department of Urology at the Columbia University Medical Center and the Herbert Irving Comprehensive Cancer Center. She has presented her research at more than fifty institutions and international conferences and has organized symposia, scientific conferences and "think tanks" for the American Association of Cancer Research, the Jackson Laboratories and the National Cancer Institute and currently serves on the Board of Scientific Advisors of the National Cancer Institute.

Xin Wei Wang joined the Institute after graduating from Shanghai First Medical College and earning his doctorate at NYU's School of Medicine. He is a world renowned liver cancer researcher, eventually becoming Chief of the Liver Carcinogenesis Section and Deputy Chief of the Laboratory of Human Carcinogenesis, Center for Cancer Research at the National Cancer Institute.

Dutch scientist Joost Verhaagen obtained his doctorate from the Rudolf Magnus Institute at the University of Utrecht before joining Frank Margolis' laboratory of Chemosensory Neurobiology at the Institute. In addition to publishing more than 200 peer reviewed papers, he has played his role in the scientific chain by mentoring 24 Ph.D students. After working in the Laboratory of Neurobiology and Behavior at Rockefeller University, he founded and runs the Laboratory for Neuroregeneration at the Netherlands Institute for Neuroscience and is strategic Professor in the Molecular Biology of Neuroregeneration at the Vrije Universiteit Amsterdam.

Pascale Léon graduated from the Institut de chimie et physique industrielles with a degree in Chemical Engineering (Lyon, France) and a Ph.D. in Life and Health Sciences from the Université de Paris VI before becoming a Roche Institute Fellow. After working in Nutley Dr. Léon eventually became the Research Program Director and later Executive Director of the Canadian Breast Cancer Research Alliance, a national group of public agencies and private foundations funding breast cancer research across Canada. In 2009 Dr. Léon led the development of a National Breast Cancer Research Framework, and in addition to authoring more than 30 publications and papers has been recognized for outstanding work in in medical research management.

Steven Herman worked in Mel DePamphills's lab at the Institute after receiving his doctorate at Tufts. Upon completing his fellowship he worked in biotech for two years before returning to Roche as a Senior Scientist and in 2004 became Director of Diagnostics Development. In that capacity he has directed the development of PCR-based in vitro diagnostic tests, including quantitative tests for HIV viral load, multiplex blood screening tests for HBV, HCV and HIV viruses, and a multiplex test for bacterial and fungal sepsis pathogens.

Peter Tattersall was one of the first postdoc fellows to work in the brand new facility, joining Aaron Shatkin's lab in 1972 to study the Replication of Animal Viruses Containing Single-Stranded RNA. He eventually went into academia, becoming a Professor of Laboratory Medicine and Genetics at Yale.

At any time an array of languages and accents resonated throughout the corridors. While some of the postdocs never really became fluent in English, they all had the ability to speak science, and the numbers have the same meaning in every language. But to celebrate the diversity each year a large international dinner was held in the large lobby of the Institute.

In addition to the postdoctoral Fellows, to insure a flow of new ideas a visiting scientists program was Instituted. As Herb Weissbach and Nat Brot wrote as the Institute developed, these Professors, "normally on one year sabbaticals from their universities, bring with them a depth of science and wisdom which adds still another dimension to the process of scientific development and discovery at the Institute. What has been happening as a result of some of these visits is that collaborative programs are initiated with Institute staff members and continue well beyond the time a Professor returns to his or her own university."

The weekly and sometimes more than once-a-week-seminars gave Institute members as well as visitors an opportunity to learn about work being done in associated fields that they otherwise would not know much about. Scientists from the corporate side of the company regularly attended these seminars, as well as presenting their research at them. Attorney George Gould described the program as "An opportunity to really rub shoulders with the elite scientists from a wide variety of scientific disciplines."

The seminar program really didn't become truly active until the mid-1970s. Prior to that Institute members were often invited to speak at other institutions around the country. But by 1976 RIMB hosted more than 100 seminars with invited speakers from around the world. In that bicentennial years invited speakers ranged from New York City's retired Chief Medical Examiner Milton Helpern to 1975 Nobel Laureate in Physiology or Medicine David Baltimore, who spoke on reverse transcription. Speakers came from as close as Rutgers and Columbia Universities and from as far away as Moscow's Institute of Molecular Biology and the

John Curtin School of Medical Research in Canberra, Australia. They spoke on subjects ranging from Kinetics of Formation of 5'-Terminal Caps in mRNA to Genetic and Molecular Analysis of Development and Behavior in Drosophila melanogaster: Y Chromosome Function During Spermatogenesis and Chemosensory Discrimination.

Among the hundreds of symposia and lectures there were a few that stood out, and among them was the Molecular Biology and Pharmacology of Proteins and Peptides symposium held in April, 1983 in honor of Sid Udenfriend's 65th birthday. More than 80 scientists from around the world, many of them having worked with or for Udenfriend participated in the two-day meeting held in Nutley, among them Nobel Prize recipients Marshal Nirenberg and Severo Ochoa. While Udenfriend was forced to step down from his post, he was so highly respected that Herb Weissbach, who assumed his position as Director, made sure that he could keep his own lab in the Institute where he continued to do productive science for the next decade.

The single most important lecture each year was the V.D. Mattia Lectureship, honoring the former Hoffmann-La Roche President and CEO, Barney Mattia, "who played a vital role in the creation" of the Institute. The award carried with it a $5000 stipend and tremendous prestige. The recipient was selected by the Scientific Advisory Board of the Institute. The first lecture, as mentioned previously, was given by future Nobel Prize winner Dr, Paul Berg. His lecture, which was held in the auditorium of the new building, was entitled Dissections and Reconstructions of a Viral Genome, and described the historic experiments conducted by his team in which duplex DNA molecules were covalently joined to one another, which eventually allowed them to insert a DNA segment containing lambda phage genes and the galactose operon of E. coli into SV40 DNA. In other words, he described to the audience experiments that were basic to the biotechnology revolution.

The winners of this award are essentially a list of the scientists who changed the world. In many cases being named the Mattia Award winner was a prelude for winning a Nobel Prize. In 1973, for example, the award was given to Sweden's Sune Bergstrom, who would win a Nobel Prize in 1982 for his discoveries concerning prostaglandins and related substances, findings that opened an important new field of biomedical research. In 1976 the recipient was 1980 Nobel Prize winner Walter Gilbert of Harvard, who developed assays to sequence the chemicals in DNA.

The co-winners in 1978 were 1990 National Medal of Science winners Stanley Cohen and Herbert Boyer, whose work laid the foundations for the biotech revolution. Stanford's Cohen presented a lecture on Site-specific Recombination and Gene Evolution while UCSF's Boyer's talk was entitled, Analysis of Plasmid DNA Replication Elements.

In 1979, for example, the recipient was César Milstein, who would be awarded the Nobel Prize in 1984, for his work that revolutionized immunology and has had wide applications in clinical biochemistry and biomedical research. The lecture he presented, entitled Monoclonal Antibodies from Hybrid Myelomas, described how he had progressed from simply trying to answer fundamental questions about the way cells make antibodies.

The 1983 recipient was MIT's Susumu Tonegawa, who would win the Nobel Prize in 1987 for his work in the field of immunology. Fittingly Tonegawa, a Japanese scientist, won his Nobel for work he had done years earlier at the Basel Institute for Immunology in Basel, Switzerland, the sister Institute of RIMB, which was founded in 1969 essentially to replicate the model in Nutley. Tonegawa, who was a student of Dulbecco, initially joined the Basel Institute as a molecular biologist but while there transitioned into immunology and carried out his landmark immunology studies. "What we have shown," he said of his work in Basel, "is that genes can change during the life of an individual, and that that change explains how we can protect ourselves against" disease-causing microbes.

The 1984 Mattia winners were Michael Brown and Joseph Goldstein who, according to the announcement, "within the span of a dozen years revolutionized our understanding of how surface receptors mediate the entry into the cell of specific ligands from the extrenal environment... Their findings have resulted in fundamental conceptual advances in our understanding of cell biology and the nature of inherited disease." They received their Nobel a year later primarily for their studies that led to the discovery and development of statin drugs.

Continuing the incredible string of future Nobel Prize recipients, Harvard's E.J. Corey received the 1985 Mattia award, a prelude to his1990 Nobel Prize in Chemistry. According to the citation he made "fundamental contributions that have resulted in the synthesis of important bioorganic compounds of great therapeutic value and the development of computer assisted approaches to organizing the logic of synthesis."

The 1986 award winner was Günter Blobel, who would win his Nobel Prize in 1999. A year later Thomas R. Cech, a world renown leader in RNA processing and chromatin structure, who would win his Nobel Prize in 1989, was honored. The first woman to win the Mattia Award was Christiane Nüsslein-Volhard in 1990, who would receive the Nobel Prize in 1995 for her work in developmental genetics. Nüsslein-Volhard, at that time the Director of Germany's prestigious Max-Planck-Institut, was another one of the pioneers of biotechnology, who "spearheaded the recent dramatic progress in our understanding of the development of the fruit fly."

Continuing the tradition of selecting future Nobel Prize winners the 1991 recipient was Leland H. Hartwell, who received his prize in 2001 for his work in discovering the protein molecules that control the division and duplication of cells.

The 1993 recipient was MIT biologist Howard Robert Horvitz, who in 2002 was awarded the Nobel Prize. By this time great strides had been made in understanding many of the processes by which DNA creates a unique being, but so much more was still to be uncovered. Among the mysteries was how the genetic information stored in DNA produces a three-dimensional organism. As Horvitz discussed in his lecture, Genetic control of cell lineage, cell signaling and cell death during *C. elegans* development. He had made great progress in this area by identifying most of the genes critical for the development of the nematode worm.

The tradition continued when biochemist James Rothman, the 1994 Mattia recipient, was awarded the 2013 Nobel Prize for being part of the team that discovered the machinery that regulated vesicle traffic, a major transport system in our cells.

For the postdocs and the graduate students and, later, even high school students learning at the Institute, this wasn't simply an opportunity to listen to giants of modern science give an address, it was a chance to actually meet these people and the other scientists who attended to honor the recipients at the reception following the speech. It exposed them to the world of their future and, without doubt, inspired many of them. For almost a decade the Institute ran an eight-week summer program for select biology teachers, allowing them to work with mentors, a Gifted Science Students program for high school students and a summer Science Honors Student Program for college undergraduates. Coincidently, a student who had studied under one of those high school science teachers later joined the Institute as a postdoctoral fellow.

The teaching program began with a group of invited students but it soon became clear that if 50 students participated the result would be 50 science educated students, but teachers could take their new knowledge back into the classroom and reach a multiple of that number. The problem was that many teachers supplemented their income with summer jobs and couldn't afford to take that time off. Roche agreed to pay the teachers $5000 to participate. The program proved extremely successful, so much so that one participant, a woman, Dr. Jean Werth, was a coauthor on a seminal paper about the discovery of a protective enzyme against oxidative damage along with Herb Weissbach and Nat Brot.

In addition to the seminars and weekly talks there also were regular journal clubs and specialty meetings. There were also prestigious industry organizations. Nat Brot and Herb Weissbach, for example, were organizers of the N.Y-N.J Molecular Biology Club, consisting of scientists from various universities, companies and research facilities in New York and New Jersey. The club, which had been in existence for many years, met regularly in the city, often at Columbia University, but it did move from place to place. Rarely did more than a dozen people attend. Finally, in 1985 Weissbach got permission from Roche to host these forums at the Institute and since some of the participants were from other pharmaceutical companies, a few of those companies helped sponsor the monthly meetings. "What started out to be a handful of people," Brot remembers proudly, "continued to grow, and by the early 1990s we were hosting catered meetings every month for as many as 150 people. Our speakers included scientific leaders from industry and some of the finest academic scientists in the metropolitan area."

343

It wasn't just a matter of giving awards, the faculty also received them. A lot of them. As the Institute gained recognition its members often were honored with recognition for their efforts. In the early years those honors were somewhat limited, consisting mostly of invitations to lecture and honorary degrees. In 1969, the first year of the Institute's existence, Syd Spector "spent three weeks as an invited Visiting Professor in the Department of Pharmacology, Stanford University." A year later, in 1970, Herb Weissbach received the American Chemical Society Enzyme Award and Gebhard Koch was awarded the Hans-Kleinschmidt Prize in Germany for work done years earlier on the Poliovirus.

Sid Udenfriend received the first major honor awarded to an RIMB member in 1971 when he was elected to the National Academy of Sciences. This proved to be important, as it gave real credibility to the newly formed Institute at a time when many people were still questioning its viability.

Udenfriend was only the first scientist at the Institute to be selected for this honor. At one time, for example, there were seven National Academy of Sciences members – four of them selected for work done in Nutley -- and a Nobel Prize winner working there; seven members of that organization out of 25-30 principle investigators is an extraordinary number. All of that combined to allow RIMB to be rated by the Science Citation Index in the late

344

1980s as one of the top five independent biomedical research Institutes in the world. The Institute clearly had achieved one of Mattia's initial goals.

The honors and the awards began coming more frequently as the prestige of the Institute continued to grow. Numerous members of the Institute were invited to serve on advisory, editorial and review boards in their areas or appear at symposiums or give lectures. They were cited for their achievements by various local and national organizations. Among numerous honors, listed in no special order and certainly not complete – and included simply to demonstrate how successful the Institute was in its effort to make an impact in the world of scientific research -- Udenfriend was elected to the Board of Governors of Israel's famed Weizmann Institute of Science, Severo Ochoa received the National Medal of Science and Aaron Shatkin became the Editor-in-Chief of *Molecular and Cellular Biology* and Editor of *Advances in Virus Research*. Anna Skalka joined the editorial board of the prestigious journal *Gene* and was appointed Vice Chairman of COGENE, The Committee on Genetic Experimentation. Claire Neurath was elected to the Council of American Society for Neurochemistry. Both Amiya Banerjee and Yasuhio Furuichi served on the editorial board of the Journal of Virology and Furuichi was Japanese Biochemical Society's 1976

Scientist of the Year. Arthur Weissbach was appointed to the editorial boards of both Nucleic Acids Research and Cancer Research as well as the Weizmann Institute's Scientific Advisory Board for Biochemistry, Michael Sherman joined the editorial boards of *Cell, Oncodevelopmental Biology and Medicine* and the *Journal of Embryology and Experimental Morphology*, Herb Weissbach was listed as one of the top 300 most cited authors (1961-1976), named Executive Editor of *Archives of Biochemistry and Biophysics* and Editor of the *International Journal of Neuropharmacology* and received the Townsend Harris Alumni Award from City College of New York and the Distinguished Alumni Award from George Washington University. He was also elected to the National Academy pf Sciences. David Webb served on several editorial boards, including J*ournal of Immunology, Lymphokines and Prostaglandins* and *Medicine*. The American Association for Cancer Research honored Tom Curran with its Rhoads Award for Meritorious Achievement in Cancer Research. Syd Spector was elected a Fellow of the American Association for the Advancement of Science and in addition to his election to the National Academy of Sciences Ron Kaback was elected a Fellow of the American Academy of Arts and Sciences. John Connor was the recipient of the Camillo Golgi Award in Neuroscience from the Fidia Foundation. Even our valued administrator, Dick Snyder, was elected President of the Society of Research Administrators 1981.

While initially forging relationships with universities was difficult, eventually RIMB members joined the faculties of colleges and universities throughout the country. Many became associated with institutions in the greater metropolitan area. Nat Brot, for example, was an Adjunct Professor at the University of Medicine and Dentistry of New Jersey and Professor of Microbiology and Immunology at Weill Medical College of Cornell University, Sidney Pestka, Sydney Spector, both Weissbach brothers, Ralph Greenspan, Tom Curran, Melvin DePamphilis and Donald Nuss became Adjunct Professors at Columbia, and Greenspan and Herb Weissbach were also Visiting Lecturers at Princeton, Paul Wasserman joined the faculty at NYU Medical School as an Adjunct Professor, and both Frank Margolis and Ron Kaback were Adjunct Professors at the alma mater of several of the Institute's founders, the City University of New York. C.A. Rosen became an Associate Professor at the Seton Hall School of Medicine. Bernard Horecker was a Visiting Professor in the Department of Molecular Biology at Albert Einstein College of Medicine; Horecker and Chun-Yen Lai joined the faculty at Cornell University Medical College as Adjunct Professors. And Aaron Shatkin became an Adjunct Professor at both Princeton and Rockefeller University.

Further away Mohammed Siddiqui was a Visiting Professor at the University of Austral de Chile, Orestes Tsolas was a Visiting

Professor at Brazil's University of Sao Paulo, Yasuhiro Furuichi was an Adjunct Professor at the University of Quebec in Montreal, Ron Kaback was the Howard Hughes Professor of Physiology at UCLA, in addition to serving as an Adjunct Professor at Columbia and Columbia Medical School and Andrew McMahon was a Professor of Biology at Harvard.

From those early days, when Udenfriend, and Herb Weissbach welcomed any university willing to form an alliance with the Institute, those relationships grew and benefitted both the learning institutions and RIMB.

The Institute had surprisingly few continuing problems. Roche fulfilled its part of the initial agreement, robustly funding the Institute without in any way interfering with its research or imposing any demands. So funding, usually the most difficult aspect of running a research Institute, was never an issue. The biggest problems were the mundane issues of daily life. There was one problem, for example, that was never adequately solved: Where were you going to live. From the inception of the Institute there just never was sufficient housing for the hundreds of people who would come to work there for two years then leave. It was, Herb Weissbach wrote, "The most pressing problem, and perhaps the most threatening, is housing. The Institute has devoted a tremendous amount of time grappling with this problem and... an adequate resolution continues to escape us."

Roche did as much as possible to find and secure safe housing for its temporary employees, but housing remained "the most critical problem to the ongoing visiting program." Initially it rented apartments on a continuous basis, with the understanding "the Institute would be free to place in the units any of its temporary staff," and would pay that rent even if the units were unoccupied. That still fulfilled only about 10% of the housing needs, leaving visiting researchers and scientists to scramble to find suitable living space. The problems got so bad, in fact, that one postdoc, according to Barbara Kerr, appeared to be living in his car. "It seemed like all of his belongings were in there," she said. "Everybody kept saying, 'Yeah, he's homeless,' but he never seemed to go home. He was always there."

Another problem that both Institute members as well as visiting faculty and researchers had to deal with was the fact that while the facility in Nutley was only 13 miles from New York City, it was just far enough away to make it difficult to reach. So even though the lure of New York was part of the attraction, but too often families with young children living on a somewhat limited budget weren't able to take advantage of that proximity. "One of the things that I think was not fully appreciated by us," Herb

Weissbach admits, somewhat sadly, "was that for some of the families the area was not ideal. For people coming to Nutley from a great distance the translocation could be difficult. Many of the young wives felt they were locked up in a New Jersey suburb where there wasn't much of interest for them."

Herb Weissbach's personal experience served as an example of the dilemma faced by other people, "My wife Renee and I had lived in Berkeley, California, which we loved, and then came back to suburban Maryland because, in my mind, the NIH was the best place to do science. But when the opportunity to be part of the new Institute was offered to me she didn't want to come up to an industrialized area in New Jersey. That was the last place she wanted to be. But she knew it was the best thing for my career and she was always, always, supportive.

"But it turned out to be more difficult for her than we had anticipated because the school system was not at the level of what the kids had in Maryland. And, in fact, that was true for several other people. We certainly should have paid more attention to the family requirements than we did. Because science is, for so many people, more than a full-time job, the spouses especially those who did not work, were left on their own a lot. In a university community, or in a large city, there often are a lot of opportunities for them. But that wasn't the case in New Jersey and the commute to New York was not that convenient. We did try our best but there really was only so much of that we could do."

Eventually though, there was one problem that the Institute couldn't overcome. And the result was the end of the dream.

Chapter Eight

A Dream Ends

By the early 1990s the list of accomplishments for the Roche Institute of Molecular Biology was both long and impressive. By any measure it was a huge success. By the most obvious measure, was it a financial success, the answer would have to be a resounding yes. While the cost of operating the Institute for almost three decades certainly would approach a billion dollars, the cumulative value of the products that came out of the Institute, including Interferon and the diagnostic products, as well as the relationship with Genentech would be far greater than those costs. The value of Genentech as an acquisition, even deducting the purchase cost, is considerably more than the cumulative operating costs of the Institute. "Without the relationship with Genentech," Ron Kuntzman said flatly, "instead of being the third largest pharmaceutical company in the world, as it is today, it probably would be thirtieth. When Sid Pestka and the people at Genentech started collaborating on Interferon a relationship was formed that changed the company forever." Added to that the profits generated by Genentech makes it an extraordinarily good deal that would never have happened if the Institute hadn't initially established the relationship. As George Gould pointed out, "It would be very difficult to look at all the patents that came out of the research, and all of the products that arose because of those patents and make any kind of true estimate of how much value there was for Roche in terms of revenue. It certainly was substantial, very substantial."

In addition to specific products, the Institute also contributed greatly to advancing scientific research. The Institute policy that allowed research that might have otherwise remained proprietary to be widely published – adhering to the historic traditions of research scientists -- undoubtedly contributed greatly to advances and discoveries made in other labs. "There is no doubt," Arthur Weissbach said, "that scientists inside the Institute made significant advances in uncovering the process by which genetic information is used is used by the cell to make proteins – especially in terms of the translation process, where information is converted from mRNA into proteins."

But many of the other accomplishments of the Institute can't be so definitively measured. According to Herb Weissbach, "There is no way to put a dollar figure on the way the Institute changed Roche. When we first presented the concept of discovering drugs through biotechnological advances, it was foreign to Roche. Roche was a chemical company. Manipulating genes and DNA to create new drugs wasn't even in their minds. They would create compounds then screen them to see if they had any physiological effect. If a compound happened to put animals to sleep they would investigate it. Pharmaceutical companies were getting insulin by isolating it from the pancreas of a pig. They were doing nothing at the molecular level. We showed Roche how to take the human interferon gene and put it into a bacteria that would produce as much interferon as you wanted. It was an entirely new approach to research. I don't think anyone would dispute the fact that the Roche of 1967, when we signed the charter, and the Roche of 1990 were different companies. The whole way of approaching research had changed. Roche was the first major pharmaceutical company to utilize these techniques and that was completely because of the Institute.

"The research carried out at the Institute…was unquestioningly at the cutting edge of molecular biology," Udenfriend said proudly. "RIMB played a key role in Roche's setting up a peptide group, as well as a biopolymer/bioengineering group for the large scale isolation of proteins. RIMB helped Roche establish a department of molecular biology with one of our senior staff members as director and a number of junior staff its initial cadre. RIMB also urged Roche to set up a central cell-culture facility, again providing the director and the space."

"In industry there always was a target," Herb Weissbach points out. "There was very little basic research going on at Roche before we came. The whole concept of using biotechnology to make drugs rather than chemistry was brand new. There is no question that we set the tone for the future of the pharmaceutical industry when we set up the Institute as a molecular biology research center in 1967. As a result of that cutting edge research the image of Roche changed drastically. Young people saw it as a desirable place to work and it made recruiting easier. Irwin Lerner said flatly, "It gave the company tremendous scientific credibility and so people were not at all reluctant to come to Roche to pursue their own careers. The Institute itself, the establishment, maintenance and sustenance of the Institute, brought more and more bright scientists from top places like the NIH to Nutley. The Institute made Roche a very desirable place for young scientists to be."

The fact that the Institute eventually was ranked by Science Watch as the fifth leading independent biological research laboratory in the world, along with legendary institutions like

Watson's Cold Spring Harbor, the Hutchinson Cancer Research Center and the Salk Institute, brought tremendous prestige to the company. "It said to the world that we weren't just a valium factory," Ron Kuntzman added. "The recognition that Roche got from the Institute certainly helped its image." In addition to Roche founding the Basel Institute, in 1970 two pharmaceutical companies located in Basel, Ciba Aktiengesellschaft and J. R. Geigy Ltd., funded the Friedrich Miescher Institute for Biomedical Research. This Institute is still in existence.

In a broader sense the creation and success of the Roche Institute blurred the previously rigid lines between academia and industry. It became the model for industry supporting the type of pure research – without specified product goals – that previously was the hallmark of academia. That was a primary reason when it was founded that so many people doubted it would last and questioned Roche's real commitment. But through the years Roche proved over and over not only its belief in the Institute, but also its respect for the charter it had signed. As Herb Weissbach wrote in his annual Director's Report in 1993, "I must again thank Hoffmann-La Roche for their continuing support of the Institute. I refer not only to financial support, but also to the adherence to the principles set forth in the Charter signed by Dr. V.D.Mattia in 1967. This has occurred through four administrations here in Nutley." It can be argued that the odd alignment of the Institute as an academic Institute operating in a corporate setting also became a model for the greater biotech industry.

The impact the Institute's training programs had on the greater world of science also is impossible to measure, but certainly is significant. More than a thousand young scientists received their early training there then took their knowledge and experience into the world. As one of them, Joe Tarnowski, says, "When I look around at what my colleagues who were there with me have accomplished there is a tremendous feeling of pride. I honestly don't think a lot of it would have been done without that experience. We learned to work hard there and love what we were doing. There was tremendous respect not just for our colleagues, but for science itself. We learned how to approach it the right way. There definitely was a legacy of excellence. As one postdoc noted, "Sometimes it's hard to believe the number of people you meet who say they were at the Institute or visited there. It's an ice breaker at times, when you run into people who matriculated through the post-doc program. You immediately know that this person has to be a professional and probably is quite capable. It's hard to imagine there was a better place to start a career in science."

"Beginning in the mid-1980s," Herb Weissbach says proudly, "if I had to give a lecture at a university or, in fact pretty much any large scientific Institute, there was a high probability someone who had been trained at the Roche Institute was located there. Our tentacles spread throughout the world – and now those people are training the next generation of scientists."

These training programs didn't benefit only the larger world of science, but an estimated 100 Institute personnel eventually moved out of the Institute into other areas of the Roche organization – some moving just a few feet across the road as Anna Skalka did -- at least temporarily, while others moved across the Pacific Ocean as Yasuhiro Furuichi did to become Head of the Molecular Genetics Department at Nippon Roche or across the Atlantic like Chris Coleclough did when he joined the Basel Institute. Most of them stayed a lot closer though, finding roles in one of the numerous departments of Roche's vast corporation. Dr. Pat Gage, for example, who had spent years at the Institute studying the molecular biology of the silkworm, eventually became Vice-President of Roche Exploratory Research – and recruited several Institute scientists for Roche's Molecular Genetics Department. And conversely, at various times researchers from the company actually set up labs inside the Institute to collaborate with members on a variety of projects or learn techniques. Even non-scientists at Roche were trained. "In fact, one of the informal courses I ran was to explain the basics of molecular biology to the Roche's lawyers. There were about a dozen lectures and the patent lawyers would sit in a room to learn about molecular biology and recombinant DNA technology. Until that time they had only dealt with organic chemicals, and suddenly they were going to have to deal with biologicals. The world was changing and we were at the forefront of it."

In his 1992 Director's Report Herb Weissbach probably came as close as anyone to summing up the impact that Institute had on both Roche and the greater scientific community, writing simply of the original concept which became an historic organization, "When you take a group of talented scientists and give them freedom and ample resources and support as we've had here at Roche, you have a formula for success."

However, there were dark clouds on the horizon and by looking at how the biotechnology revolution evolved one could almost predict what the fate of the Roche Institute would be.

Within only a few years of the founding of Genentech, the first true biotech company, the business of science had exploded. Beginning with the ability to produce human growth hormone and human insulin by recombinant DNA technology in the 1970's the possibility of producing other products through genetic engineering had become the commercial pot of gold waiting at the end of a rainbow. This was the advent of the billion dollar molecule. By the early 1990s it was estimated there were more than a thousand biotech start-ups, which were defined loosely as companies using biotechnological processes to solve problems and create useful products in a variety of fields including agriculture, manufacturing and medicine. While once disdained by purists in the scientific community, eventually these companies began employing prestigious scientists, saw results published in respected journals, and made some very important – and patentable – discoveries into the workings of the world. Biotech had become a multi-billion dollar industry and played a growingly important role in the world economy – in addition to its impact on science.

Eventually though, it became apparent to investors that very few companies would ever duplicate the extraordinary success of Genentech. On average, it took more than a decade and required an investment of hundreds of millions of dollars to bring a single drug to the market. Because of that, a lot of these companies joined forces with major pharmaceutical companies, who had the finances to support their research, the experience in dealing with governmental agencies through the long regulatory process and the patience to accept the need for years of research and development – and the high potential for failure. And among the biotech capabilities that made these relationships very enticing for Big Pharma was the ability to add targeted drugs to the relatively new phenomenon of the block buster drug.

While in past decades a drug company would of necessity rely on a variety of profitable products to produce its annual profits, suddenly one drug, one multi-billion dollar product, could support an entire company for years – just as valium had done for Hoffmann-La Roche. But in addition, biotechnology made it possible to create drugs targeted at a relatively small number of patients suffering from a disease, which could be marketed much less expensively to a specific population. According to an industry study, "The biotech firms provide a 'farm system' for the big drug companies, doing the early research and development, while the big drug companies invest in promising late-stage development and provide the capacity to market and distribute the drug once it has been approved."

As a result by the mid-1990s a majority of the drugs either approved or in clinical trials had been created by biotech companies – and the pharmaceutical industry funded almost two-thirds of those companies.

According to attorney George Gould that led to a major change in philosophy throughout the industry, "The whole idea of cherry picking had become the rage in the pharmaceutical industry. Why support a huge research effort when you could wait until somebody developed a product to the point of at least demonstrating a proof of principle in the clinic, and then buy the product, or buy the company, which saved a lot of money that had traditionally been invested in basic research."

Roche's deal with Genentech served as a high-profile example of the relationships being formed that brought the two industries together. But it also meant that fundamental changes were taking place through the pharmaceutical industry.

The independence of the Institute, which had long been a source of great pride for its corporate parent, was becoming increasingly costly in a growingly competitive – and bottom-line oriented – biotech industry. The continued excellence of the Institute was never an issue, but for the first time there was some question about how it fit into the changing world. As Jürgen Drews who joined Roche in the late 1980s as President of International Research and Development reported, "An even more rigorous innovation strategy resulting in an even greater proportion of valuable drugs is the call of the day. The Roche Institute of Molecular Biology is an instrument in this innovation strategy. But what is the scientific mission at a time when molecular biology has become the mainstream of all biomedical science and when the selection of a few major topics and approaches, perhaps even the pursuit of one dominating idea, has become mandatory for eventual success?

"We – the Institute scientists, our friends in the scientific community and research management at Roche – will have to find an answer to that question." But then Drews concluded, ominously, "The research-based pharmaceutical industry has come to a critical point in its history. While the number and quality of exploitable scientific opportunities are unprecedented, our ability to address these opportunities will be increasingly limited by economic constraints. We will have to make difficult choices."

The Institute actually was positioned to find the elusive answer. In the decade after Weissbach had replaced Udenfriend he had successfully implemented a new structure while supervising the turnover from the original scientists to a younger staff. He also had intentionally recruited scientists with expertise in the developing – and intellectually hot -- areas of research to Nutley. He brought in several respected scientists working in areas such as signal transduction, gene expression and regulation, neuroscience and embryology, as well as appointing Tom Curran, formerly the Head of the Department of Molecular Oncology and

Virology, as Associate Director. While in reassuring ways all of those elements that had been developed through decades of success and made the Roche Institute a very special place to do research had remained in place, in fact it had been injected with new and younger energy with an emphasis on those most promising areas of research and development.

And then everything changed.

As George Gould points out, the Institute was no longer as unique as it was once was; "Those early technological advances that the Institute had helped pioneer; protein purification, genetic manipulation, had become somewhat commonplace, partially because the young scientists that had trained there had left and spread that knowledge."

To better understand this wave that had swept over the industry in 1991, according the Ron Kuntzman, then the Vice President in charge of Research and Development in Nutley, Roche hired a major international consulting firm to analyze its corporate research structure. As he remembers, "Irwin Lerner told all of us on the Executive Committee that this company was going to be examining the research capabilities in Switzerland and that they would not be looking at Nutley. But little by little it became clear that Lerner was not being told the truth, as it became obvious that Nutley was going to be involved in this study.

"The conclusion was that the facilities in Basel and Nutley had to be brought together as one research organization." At that time Kuntzman was running R&D in America while Jürgen Drews was his counterpart in Europe. Those two men would usually meet twice annually to discuss the ongoing research being done at their respective Institutes, "What we were working on, what we weren't working on; what we should be working on, what we shouldn't be working on. It had worked quite well for several years, but when this consulting firm urged that all research be put together in a single place, Drews was appointed the Head of Research and Development worldwide. For the first time these two great organizations, Nutley and Basel, were going to function together. Drews moved to the United States, and immediately made it clear to everyone that he was in charge of Roche research. While initially Drews may not have been considering closing the Institute, Kuntzman added, soon after he settled in New Jersey he began talking about moving the Institute to California.

Well before any final decision about the Institute's future was made many of the senior staff, the people who had successfully established the Institute as one of the best research centers in the world, had already left. They were in high demand.

Anna Skalka had been among the first to leave, accepting the position of Scientific Director of the Institute for Cancer Research and Senior Vice President of Basic Science at Philadelphia's Fox Chase Cancer Center. "It was time for me to go," she says. "A lot of people tended to stay for a long period of time. It was a happy place. But I was offered a bigger and more important position. I felt there was more I could contribute there. I also began to feel that maybe the Institute needed new energy. It's always sad to leave a place you enjoy, but in the world of science people do disperse but manage to keep in contact."

Aaron Shatkin left at about the same time and accepted the position of founding Director of the new Center for Advanced Biotechnology and Medicine, a research laboratory jointly administered by University of New Jersey and the University of Medicine and Dentistry of New Jersey. He spent the rest of his career there, personally building the CABM into a world class facility while continuing his own research into deciphering the mechanisms of eukaryotic gene expression, with special focus on viral and cellular transcription and translational processes.

Severo Ochoa left the Institute in 1985 to return to Madrid in celebration of his 80th birthday. As Herb Weissbach wrote in his annual Director's report, "Severo has been with the Institute since 1974 and his leaving creates a void that is impossible to fill...

Though his home base will not be in Nutley, he has been appointed a Visiting Scientist which will permit him to maintain scientific and personal interactions with the Institute staff in the future."

After serving as the Associate Director of the Institute, Arthur Weissbach had retired in 1989, and after spending one additional year as an Adjunct scientist, moved to Sanibel, Florida, eventually becoming a visiting scientist at the Sanibel-Captiva Conservation Foundation's Marine Laboratory since 2004. In addition to working on delineating and defining fluorescent patterns in red tide organisms (HABs) he also began studying the Dengue virus as a scientist in the Biotechnical Division of Florida Gulf Coast University.

Bernie Horecker, who had been elected to the National Academy of Sciences before coming to the Institute, took a position as Dean of the Weill Cornell Graduate School of Medical Sciences

Sidney Pestka had left Nutley in 1986 to become Professor and Chairman of the Department of Molecular Genetics, Microbiology and Immunology at UMDNJ-Robert Wood Johnson Medical School in Piscataway, New Jersey. In 2006 he was named the recipient of Lemelson-MIT Lifetime Achievement award, for his "tremendous contributions to the biotechnology, chromatography and pharmaceutical industries, and for improving the health of those suffering many serious diseases, largely for his groundbreaking work at the Institute, developing antiviral treatments for chronic hepatitis B and C, multiple sclerosis and cancers.

Another early member, Ron Kaback, had left in 1989 to become an Investigator of the Howard Hughes Medical Institute and Professor of Physiology and Microbiology, Immunology & Molecular Genetics, as well as a member of the Molecular Biology Institute at the University of California Los Angeles. In addition to being elected to the National Academy of Sciences among the honors he received for his work in Nutley were the Selman A. Waksman Honorary Lectureship Award by the American Society of Microbiology, the Lewis Rosenstiel Award by Brandies University and the Kenneth Cole Award by the American Biophysical Society.

While Syd Spector left Nutley in 1990 to join the faculty at Vanderbilt University – one of the last remaining charter members, he continued his affiliation with the Institute. By that time the work he had done at the Institute had led to his election in 1987 as President of the American Society for Pharmacology and Experimental Therapeutics.

Craig Rosen left in 1992 to go into private industry, co-founding and becoming the first senior manager at Human Genome Sciences. He assumed responsibility for directing Human Genome Sciences' research and development efforts and eventually rose to CEO and President of that multi-billion dollar company.

Ralph Greenspan also left in 1992 but he went into academia, accepting a position as a Professor of Biology and Neural Science, and Head of the W.M. Keck Laboratory of Molecular Neurobiology, New York University. After moving to San Diego where he became a Senior Fellow in Experimental Neurobiology at the Neurosciences Institute, he eventually became the Associate Director of San Diego's Kavli Institute for Brain and Mind.

This turnover in the Institute staff made it possible for Herb Weissbach to bring in younger scientists in new research fields. However, the fate of the Institute was further compounded in 1993 when Herb Weissbach, having completed the reorganizing and revamping of the Institute, and believing it to be primed to move into the next century, decided to step down as Director. "I was approaching 65 and I told Roche that it was time to replace me with a younger scientist. I was also deeply concerned at that time, because Jurgen Drews was the first Roche Research Director who wanted to determine what research areas the Institute should be concentrating on. Previously Roche management had honored the Institute Charter and had never directly interfered with the research in the Institute.

After Weissbach announced his intention Roche began searching for a new Institute Director. They put together a top-flight scientific advisory board, which was chaired by molecular biologist Leroy Hood, to examine all the alternatives and suggest potential candidates for the position of director. But they also began thinking about moving the Institute out to California to be closer to Genentech. There was considerable discussion about moving into a new area, called genomics, which was becoming more important. They saw this as an opportunity to plan for the future, and as far as Jurgen Drews was concerned, the future was genomics.

Genomics was a term coined in 1986 by geneticist Tom Roderick at a meeting concerning mapping the human genome. Essentially it meant looking at genetic information from an organism to discover the differences in the DNA of health and diseased bodies, enabling scientists to design a specific treatment. It was a means of tailoring the treatment to the individual organism. From the very beginning people believed this was going to become an important therapeutic field. And, for a time at least,

370

it seemed like all the new advances in treatment were going to come from this area. Jurgen Drews, in particular, wanted to refocus the Institute on the genetic basis of human disease."

Weissbach remained confident in his belief that "Roche really wanted the Institute to continue, even if it wasn't in Nutley. At one point they actually made an announcement that the Institute would be relocated to California. Roche even had announced a new name for it; the Roche Institute of Molecular Biology, once the cutting edge of science, was to become the Roche Genomics Institute in Palo Alto. They interviewed several top scientists to replace Weissbach. Among them were Tom Caskey, who had been the Head of Medical Genetics in the National Heart and Lung Institute, before moving to the University of Texas. Caskey instead accepted a position as Senior Vice President for Research and Trustee and President of The Merck Genome Research Institute at the Merck Research Laboratories. They brought in Eric Lander, the MIT Professor who had founded the Whitehead Institute/MIT Center for Genome Research. Under his leadership the Whitehead Institute had become one of the world's leading centers of genome research. Lander considered the offer but he wanted to run the Institute while continuing both teaching at MIT and running his Institute, which obviously wasn't possible. They also considered Dr. Jim Rothman, who was a researcher at

Memorial Sloan Kettering Hospital in New York City, as well as a Member of the National Advisory Council for Human Genome Research at the NIH, but it wasn't a good fit for him. Roche was looking, but it was proving very difficult to find the right person."

"It was becoming more and more obvious that Roche didn't really know what to do with the Institute," George Gould explained. "It also was clear that the interest in supporting a pure research Institute was waning. Money was an issue too. The Institute was costly to operate and there was no way of measuring the return in dollars and cents. There were some people in management who thought that maybe if the Institute was moved to a more suitable place they might be able to interest other companies in offering some support. With all the biotech start-ups and venture capitalists on the west coast that would be the right place, they thought.

"One of the first ideas was to form a sort of partnership with Stanford University. I was tasked with approaching Stanford University, along with some people in management, to see if we could interest them in establishing an Institute there. The plan was to transfer some of the scientists from the Nutley campus to Stanford in hopes of attracting funding from multiple sources, including Genentech which was not yet completely owned by Roche. We actively explored that possibility; Stanford certainly showed some interest and it would have been an important addition for them. We got to the point of preparing a general term sheet. But eventually the Stanford initiative collapsed."

Apparently Roche hadn't reached any final decisions. The significant question appeared to be where the revamped Institute would be located. The probability that it would be moved to California became far more likely in the fall of 1994 when Roche purchased Syntex, a California pharmaceutical company for $5.3 billion. As Sid Udenfriend pointed out, "The acquisition of Syntex gave management the option of setting up a research facility in Palo Alto, California, an area that would... permit them to redirect its research to genomics more rapidly, rather than overhauling the existing RIMB."

Several of the candidates for the position of director had indicated that moving it to the west coast would make the position far more desirable. Michael Steinmetz, Roche's Vice President for preclinical research and development, told a reporter, "They did not have particular problems with the current site, but thought if the Institute would be in the San Francisco area or the Boston area they would be able to benefit from a number of high-quality academic institutions with which they could collaborate. "So based on this kind of feedback and based on the fact that Roche was suddenly in a position to move the Institute to the West Coast, the company decided that that is what we would do. Once that decision was taken, we had several people call us back and tell us, given the new situation, they would actually now be interested in applying again for the position."

Had they been able to find the right person to replace Weissbach it is possible the Institute could have been moved successfully to Palo Alto and enjoyed many additional years of success. But for a variety of reasons they were never able to fill that position. The lack of new leadership to guide the Institute into the future, the indecision about where it should be located, the reluctance to continue to commit substantial funds to undirected research and the existence of countless start-ups created to be sold all combined to force Roche management to make the almost inevitable decision.

Weissbach and Udenfriend were aware of the various attempts to resolve the problems. There had been little, if any indication that Roche management was suddenly going to reach that unexpected decision – which came as a surprise. In fact, a week before Drews made the announcement Weissbach had been with him at the dedication of a new $100 million multidisciplinary science building on the Nutley grounds and he hadn't hinted at the forthcoming announcement. "One day I was called into Drews office," Herb Weissbach remembers. "He told me the search was finished. I thought by that he meant they had found my replacement. That wasn't it at all. He told me that a decision had been made to close the Institute. The genomic research was going to take place at Syntex – and they probably weren't going to move anybody from Nutley to Palo Alto. I was shocked. Everyone was shocked."

It was, according to RIMB scientist Robert Margolskee, "shocking, surprising, depressing, disruptive."

Any doubt that it was real was dissipated by a memo Drews circulated in early November 1994, in which he wrote that Roche had decided to eliminate 540 jobs on the Nutley campus, including the Institute staff. He wrote, "We plan to relocate" the Institute at Syntex in Palo Alto, and "rebuild it in the new location with a strong scientific focus on genomic sciences as they relate to multifactorial diseases."

According to Udenfriend, "Once it was decided to set up an Institute in California, RIMB became a casualty of the company's downsizing." As he added later, with sadness, "Continued support of RIMB by Roche was generous to an extent that, again, has not be seen in academia or government...I was probably naïve to believe that any company could support an Institute such as RIMB indefinitely." For both Udenfriend and Weissbach this was the beginning of what would be, as they later admitted, the most difficult two years of their careers.

Roche management offered several suggestions why it had decided to close the research center, but Roger Longman, the editor of the healthcare industry newsletter In Vivo, spoke for many people in the industry when he said that – after three decades of complete independence -- the Institute had come to be judged by company management on the basis of its direct contribution to Hoffmann-La Roche products, "The Institute was a luxury that Hoffmann-La Roche couldn't afford... Everybody in the drug industry is looking to cut costs and make fewer assets produce more...

"The real impetus for moving the Institute to Palo Alto – and its anticipated reinvention as a center of genomics research – was, according to several scientists inside and outside the Institute, the perceptions of Jurgen Drews... Apparently Drews concluded that the only way to reshape the Institute as a leader in human genome studies was to start all over with new scientists who are more involved in gene mapping and sequencing."

The scientific community also was upset. The Institute was still highly respected. Just a few years previous it had been rated one of the top five independent biomedical research Institutes in the world. As an article in Science reported, "Scientists at RIMB in Nutley, New Jersey were shocked to learn this week that a major shake-up will close their renowned research center... The Institute...has made significant contributions to basic science, including the cloning of interferon, pioneering work in neuroscience and groundbreaking work on human taste sensation and the regulation of plant fungi. 'The place is in turmoil,' said one widely published staffer."

While for almost three decades Roche had lived up completely to its obligations, this decision cast it in a harshly negative light. A group of 16 of America's most respected scientists, among them five Nobel Prize winners and several current or former senior Institute researchers, sent a scorching letter to the respected scientific magazine, *Nature*, which was published in January, 1995. "We view these events with great sadness," they wrote, "most importantly because they occur at a time when support for basic research is dwindling. The RIMB has been unique since its inception in 1967, when it became the first institution devoted entirely to basic research supported by a pharmaceutical company... The RIMB is unique and has succeeded in its mission in research and training to an extent that has succeeded expectations. As our understanding of the molecular biology of cellular processes is far from exhausted, it is difficult to understand

Hoffmann-La Roche's decision to close the RIMB in Nutley... For the larger scientific community, it represents yet another loss in support for the kind of fundamental research that is essential for scientific progress but for which there is no vocal public constituency."

There was considerable grumbling. Associate Director Tom Curran believed Roche was making a fundamental and potentially very costly mistake, pointing out, "It's dangerous to focus on technologies per se, rather than interesting and important biological questions... The Institute was a fantastic place. For more than 10 years I had complete intellectual freedom. I was able to do science...I was able to change direction when the thought occurred to me. I won't be able to do that someplace else." The Administrative Director, George Cardinale pointed out, "The predominant feeling is one of sadness that this noble experiment is over. No one believes (the new Institute) will be as free as this one."

As if to emphasize Curran's point, an article in *Popular Science* that appeared at about this same time demonstrated how that freedom sometimes resulted in completely unexpected discoveries. A chestnut tree fungus had killed literally billions of trees since it was introduced to America at the turn of the 20th century. While it killed the tree, this blight did not kill the below-ground shoots. "A promising treatment developed by two researchers at the Roche Institute of Molecular Biology could enable the stumps to grow up again. Gil H. Choi and Donald Nuss have produced synthetic viruses that convert the deadly blight fungus to a more benign form. Spores from the converted fungus could be sprayed on infected stumps. The altered fungus could breed with the virulent form, rendering it benign... The chestnut treatment has broad implications, says Nuss, 'It's a model system to demonstrate the concept of engineering a virus to control fungal diseases,' including potentially, Dutch Elm Disease." And while this treatment never proved effective in the field it did serve to reinforce the fact that what really was being lost were the endless possibilities that resulted from giving talented scientists the freedom to do what they do best.

There actually was no way to fight for the survival of the Institute. Roche management had every right to make that decision; they had created and funded the Institute, spending tens of millions of dollars supporting the highest traditions of science: funding scientists to follow their own curiosity, without any restrictions, and in most cases sharing the result of that work with the entire scientific community. Even at that moment, even with the surprise and disappointment and frustration and anger, it was impossible to overlook the amazing contribution Roche had made completely voluntarily. And in an increasingly difficult and competitive marketplace it was truly astonishing it had survived so long.

But for people like Herb Weissbach and Sid Udenfriend this was as much a personal as professional loss. They had been there when the Institute was nothing more than a dream. With the help of Art Weissbach they had negotiated the charter, hired the staff and created the culture that made the Institute a center of excellence. They had been there through every success and had seen the Institute develop an international reputation. Herb Weissbach, as Director at that time the closing was announced, was most significantly affected. Was he responsible in part for the decision Roche made? This haunted him and as the negotiations with Roche management began he was determined to make sure that everyone in the Institute would have a position to go to. He knew he had to obtain severance packages, especially for the members, to make these forced moves as easy as possible. Roche did assist in the process. Weissbach had numerous meetings to work out all the details, but Roche actually proved amenable. As Udenfriend wrote, "The current dismantling of the Institute is painful. Fortunately, the pain has been lessened for two reasons. First, our scientists find themselves in high demand, and most have received excellent job offers. What minimizes their pain further and adds to their attractiveness to outside institutions is the generous policy that Roche has adopted toward the termination of the Institute's professional staff."

The details of this policy were worked out primarily through negotiations among Herb Weissbach, Jurgen Drews, President of Hoffmann-La Roche's R&D worldwide and the legal team at Roche. Scientists received personal severance packages proportional to their years in the Institute as well as attractive bridging grants toward the start-up of research at their new laboratories. They were also permitted to take all their equipment, including large instruments, computers and furniture, with Roche paying moving expenses."

Those concessions made Herb Weissbach's task of finding a soft landing place for every scientist considerably easier. "These were very good scientists who were very attractive to universities; in addition to their experience they were coming with some start-up money, and with equipment. For universities that had the space, offering them a position was a no-brainer. A lot of them took their whole group with them when they moved."

When the announcement was made that the Institute was going to close Weissbach and Udenfriend, the men who had planned the building, vowed to make sure everyone had a place to move to. Unfortunately, Sid Udenfriend and his wife Shirley were in a serious car accident and the bulk of the responsibility to make sure that everyone ended up in a satisfying position was left to Weissbach. Unlike the initial members, many of the scientists were young with years of productive research ahead of them. A significant number of them were placed comfortably into senior positions. As examples, Tom Curran, who had moved from Roche research to the Institute and eventually became Associate Director, accepted a position as a Member and founding Chairman of the Department of Developmental Neurobiology at St. Jude Children's Research Hospital in Memphis, Tennessee. In 2001 he became President of the American Association of Cancer Research in 2001 and served on the Board of Scientific Advisors of the National Cancer Institute from 2000-2005. In 2005 he was elected to the Royal Society and a year later accepted the position of Deputy Scientific Director of Children's Hospital of Philadelphia and Professor of Pathology and Laboratory Medicine.

Institute Member Don Nuss became Director of the Center for Agricultural Biotechnology -- currently the Center for Biosystems Research -- and Professor at the University of Maryland Biotechnology Institute. Thomas Sato accepted a position as an Assistant Professor at Harvard Medical School and later went on to become a Professor of biology at the Nara

Institute of Science and Technology, Japan, Professor at Cornell University, United States and Centenary Institute in Sydney, Australia. After a decade as an Institute Laboratory Head, Mel DePamphilis moved to the National Institute of Child Health & Human Development, Bethesda, MD as a Section Chief and member of the Senior Biomedical Research Service. When Christine Rushlow closed her own lab at the Institute she accepted an offer from NYU Langhorn Medical Center where she eventually became a Full Professor in the Department of Biology. And Herbert Ennis became a Senior Research Scientist in the Department of Anatomy and Cell Biology at Columbia University's Medical School.

Nathan Nelson returned to his native Israel, setting up his lab in the Biochemistry Department at Tel Aviv University. While there he was instrumental in starting the The Daniella Rich Institute for Structural Biology and became its Director. Eventually he served as a Vice Chairman of the International Institute of Molecular and Cell Biology in Warsaw founded by UNESCO, as well as President of the Israel Society for Biochemistry. And in 2013 received the Israel Prize for Life Sciences.

Paul Wassarman, who had been Chairman of the Cell and Development Biology department, moved his lab to the Icahn School of Medicine at Mount Sinai where he was given the Lillian and Henry M. Stratton Professorial Chair of the Dept. Molecular,

Cell, and Developmental Biology. Robert Margolskee also joined Mt. Sinai's Icahn Medical School, as an Adjunct Professor in the Department of Physiology and Biophysics, eventually leaving to become Director and President of the Monell Chemical Senses Center, a non-profit research center in Philadelphia.

James Ofengand had spent most of his professional career at Nutley, joining the Institute in 1969 as one of the first Associate Members and eventually became a full member. It was there he did his important work on transfer RNA, creating the building blocks for later research and discoveries. He left in 1996 to become a Research Professor in biochemistry and molecular biology at the University Miami School of Medicine.

Michael Newman, who had been focusing on cancer fighting drugs at RIMB, moved to Novartis as executive director of Oncology and head of Biology for all U.S.-based oncology research before consulting with several start-ups then founding his own company, Decoy Biosystems in San Diego, where he has continued to do his research in cancer immunotherapy.

Colin Stewart accepted a position at the National Cancer Institute at the NIH as Chief of Laboratory for Cancer and Developmental Biology. After more than a decade in that role he joined the Institute of Medical Biology (IMB) in Singapore under the Agency for Science, Technology and Research as a Principal Investigator.

James Morgan joined Tom Curran at St. Jude's Children's Research Hospital in Memphis as the Co-chairman of the Department of Developmental Neurobiology.

Jean Gautier accepted a position at Columbia University where he eventually became a Professor in the Genetics and Development Program at the Herbert Irving Comprehensive cancer Center, Paul Lieberman joined the Wistar Institute in Philadelphia, a designated National Cancer Institute Research Center, Tom Gridley joined the independent Jackson Laboratory in Bar Harbor, Maine, where he continued his genomic research in animal models of disease, and Bai Lu moved to Bethesda, Maryland where he became a Section Chief at the National Institute of Child Health and Human Development at the NIH, eventually returning to China where he was appointed a Vice President in Biology at GlaxoSmithKline and then became a Professor and Executive Vice Dean, School of Medicine at Tsinghua University in Beijing, China.

Nat Brot, who had been at the Institute from the very beginning, when it still was just a dream and sketches on paper, accepted a position as a senior scientist at the Hospital for Special

Surgery in New York and Professor of Microbiology and Immunology at the Weill Medical College of Cornell University. He spoke for many people when he talked about leaving the Institute. "I left there with bittersweet feelings. It was a mixture of sadness for the end but pride – even joy -- in our accomplishments. It was like losing an old friend, but when I looked back on my productivity I had great satisfaction. When we began planning the Institute almost thirty years earlier many of us simply hoped that we could last longer than some of the other private research institutions, which was about five years. Not only had we done that, we had managed to become a world renowned scientific institution. It was simple; if you couldn't do research at the Institute, then you can't do it anywhere in the world. All you had to do was show up and work."

Nancy Woychik accepted a position in the Biochemistry Department at the UMDNJ Robert Wood Johnson Medical School in Piscataway, New Jersey and Steve Dalton, went back to Australia for a while but came back to the United States where he is a Professor at the University of Georgia.

It took a bit of time before everyone was settled in new positions, but the lessons taught and learned at Nutley continued to make a meaningful impact on the world. The people who had been there carried with them both pride and spirit about what had been accomplished there. The actual closing of the lab took almost two years as one by one individual labs were closed and people left. The once noisy corridors, alive with the excitement of discovery, were quiet. Science magazine reported, "On a recent visit to the Nutley facility, some laboratories functioned normally, others, where lab heads had already left to take new jobs, were dark and empty. A skeletal support staff remains, as secretaries and technicians have been leaving whenever they can find jobs."

The last two people to relocate were Herb Weissbach and Udenfriend. Udenfriend accepted a position as the Director of the Drew University Charles A. Dana Research Institute for Scientists Emeriti, an Institute that had been established to allow many of the retired best scientists in New Jersey to offer guidance and their expertise to undergraduate students at the university. He and his wife Shirley then moved to Atlanta to be close to his daughter.

Initially Weissbach considered a position at the University of Miami Medical School. But an initial trip to the Miami campus turned out to be more than a two-hour drive from his Florida winter home. He would have to move closer to Miami Medical school. "My wife, Renee, had already sacrificed so much for my career. She loved our place in Florida and I could envision the tears in her eyes if I asked her to move again," he remembered. "I just couldn't do it. But as I was driving home I saw a small sign for

Florida Atlantic University. Florida Atlantic University? I didn't know anything about it. I went home and wrote a letter to the Chair of Biology." Within days Weissbach met with John Wiesenfeld, the Dean of the College of Science. Wiesenfeld, a chemist by training, was a New Yorker who had attended the Bronx High School of Science and the City College of New York. FAU had very little modern science in its curriculum and Wiesenfeld asked Weissbach to establish a Center for Molecular Biology and Biotechnology. It turned out to be a perfect landing spot for Weissbach, a position that "presented an opportunity to build a research program in molecular biology and functional genomics from scratch."

Herb Weissbach had vowed to be the person who shut out the lights for the final time. By mid-1996 he was the only senior faculty left and his lab equipment had been moved out of the Institute building to the main research building on the Roche campus. Gradually the Institute building had been emptied of both people and equipment. The final day at Roche for Herb Weissbach was in early December, 1996. On a dark and dreary Saturday he was alone in the research building as the last of his instruments were being loaded onto a moving van. The place was deadly quiet - - and then he heard a tapping noise that sounded like someone walking with a cane. The tapping sound got louder and suddenly

Sid Udenfriend appeared, a wry smile on his face. The two men had not seen each other for months because of Udenfriend's car accident. Weissbach found a couple of old chairs to sit down and asked "what are you doing here?"

Sid replied, "I found out that this would be your last day, the end of the Roche Institute. Herb, we started it together and I wanted to be here with you when it ended"

The two men spoke quietly until the final piece of equipment was removed, then reluctantly rode down to the lobby on the freight elevator. Both of them had tears in their eyes as they shook hands for the last time. Then they walked to their cars and drove out of the Roche complex, Weissbach intentionally not looking back, taking with them three decades of memories.

That was the last time Herb saw Sid. In 1999 Sid Udenfriend passed away in Atlanta after undergoing surgery. Herb was the only Roche Institute scientist at a small grave site ceremony, but the following year Weissbach arranged for a memorial event at Drew University, attended by many of the scientists Udenfriend had influenced, to celebrate his life.

Epilogue

By the time the Institute closed its doors for the final time the way science is done had taken another historic step forward in ways those early scientists, working in a room in their own homes, could never have envisioned. Not surprisingly, in 2000 Roche also closed RIMB's sister organization, the Basel Institute of Immunology in Switzerland, which had been home for three Nobel Prize winners. Management announced they would be transforming the campus into a medical genomics center. In 2012 the decision was made to shut down the Nutley campus completely. As Roche Chief Executive Officer Severin Schwan wrote in an e-mail to all employees, "By far the toughest aspect of this restructuring is the decision to close our U.S. site in Nutley, which is not only an important company location, but a part of our history. But given the challenging environment we face and our own growth prospects, we need to free up resources that we can invest in the most promising projects with the highest likelihood of success."

A year later they began knocking down the buildings and by December, 2014 only five of the 40 buildings that had once stood on the site remained. One of those five buildings was identified as #102, 'the former Roche Institute of Molecular Biology, the place where Interferon was developed.' While attempting to sell it, agents described the 176,000 square foot building – which was redesigned in 2011 -- as being in 'move in' condition, having biology and chemistry laboratories, as well as an 85-seat auditorium.

The demise of the Institute turned out to be a beginning, not an end. Slightly more than a decade later, in 2009, Roche moved its corporate headquarters from Nutley to San Francisco, close to the Genentech campus. The relationship with Genentech is unquestionably the major long term impact that the Institute has had on Roche. This never could have been predicted from what Udenfriend, and John Burns had cooked up at a cocktail party. The final closing of the Nutley campus in 2014 represented the end of a relationship between Roche and Nutley, New Jersey that had begun in 1929, when its Art Deco building was opened. In 2015 it was announced that the Nutley site would become the home to a new medical school formed in partnership between Hackensack Meridian Health and Seton Hall University. In 2016 the 116 acre Roche Nutley site was sold to PB Nutclif 1 LLC, an affiliate of Prism

Capital Partners LLC, who will lease part of the site to the new Medical complex.

Ironically, oh so ironically, Professor Erik Gordon of the University of Michigan's business school explained the move to California had become necessary because "The new innovation in biotech, in genomics is happening elsewhere. It's happening in places where there are graduate educational institutions that have research faculty doing that, and New Jersey really doesn't have that." Corporate officials also acknowledged that the move was part of a larger reorganization in which the research capabilities were being consolidated to save money.

And as Roche settles into its relatively new site in San Francisco, the memories of what took place for a brief span in Nutley will continue to reverberate in science for decades. As Sid Udenfriend described it succinctly, "For more than a quarter of a century we operated a research center that was the envy of scientists throughout the world, a Camelot of the biomedical sciences."

Acknowledgement

Herbert Weissbach: Working with David Fisher was both a learning experience and a pleasure. We formed a special bond that is here to stay.   Thanks to my brother Art and Nat Brot for their unwavering support in having the story of the Roche Institute told. I also would like to acknowledge as a group the former staff at the RIMB, and others at Roche and elsewhere, who were interviewed and contributed valuable insight and information. However, some individuals deserve special mention for their knowledge filled important gaps that could not have been obtained elsewhere. These people include Irwin Lerner, George Gould and David Goeddel. But because the number of people contacted was relatively small, I especially wanted to include a list, that is at the back of the book, of all the staff and people associated with the Roche Institute, since everyone contributed in some way to the image and success of the Institute. Also special thanks to other people not mentioned in the book including Alexander L. Bieri,

393

Curator and Lionel Loew from The Roche Historical Collection and Archive, at Hoffmann-La Roche in Basel and Ivan Lefkovitz , a founding member of the Basel Institute of Immunology for their advice and assistance. Thanks also to Ms. Marjorie Cazeau of Florida Atlantic University, for her help in sorting important information from the RIMB annual reports and other tasks involved in preparing the book. Special acknowledgement to Joe Verola who took all of the individual pieces that we provided him, and he magically turned them into a book.

Last, but far from least, I want to thank my wife Renee for her advice, patience and support during this project and being my loving partner for life.

Arthur Weissbach: I want to express my deep appreciation to Annette Weissbach, my daughter in-law, for her help and support.

Nat Brot: I would like to express my love and appreciation to my wife Gloria for believing in me and who was very instrumental in my life and career.

David Fisher: Each project brings with it opportunities to learn and explore. Telling the history of the Roche Institute brought me into the fascinating world of experimental science. I would like to begin by acknowledging the extraordinary dedication of Herb Weissbach, who wanted this story told and did everything possible to make it happen. His perseverance and his understanding – in addition to his friendship – are appreciated far more than he realizes. I also would like to offer my appreciation to my friend Joe Tarnowski, who introduced me to the Roche Institute and Dr. Weissbach. I was fortunate to spend time with Arthur Weissbach and Nat Brot, whose unfailingly good nature as well as their dedication to the Institute made this book possible.

During my research I conducted numerous interviews which opened this world to me; what struck me is how much all of these people I spoke with – without exception -- have such fond memories of the time they spent there and the enduring relationships they made. I heard a lot of love as I conducted these interviews and I greatly appreciate the time each of you gave me.

All of those interviews were transcribed by Casson Masters' Scribecorps, and as I have said, I could not possibly have done this work without them.

I also would like to acknowledge the invaluable work done by Joseph Verola to turn these words into a book. I truly value his advice and our long friendship.

Finally, a theme I heard continually as I worked on this project was how important the support of the wives was to the establishment and success of the Institute. That was a different time and different contributions were expected of women, and it is clear that RIMB could not have succeeded without them. That, of course, got me thinking about my own wife, Laura, and how she makes possible everything that I do. To say simply I appreciate her is a great understatement. I value her as my partner, my greatest supporter, my best friend and the woman I love.

List of People Associated with the RIMB

Members

Albert,Vivian   R.Ph.D;   Banerjee,Amiya   Ph.D; Berkowitz,Barry,Ph.D; Bick, Michael,Ph.D; Blume, Arthur, Ph.D; Boublik, Miloslav, Ph.D; Brot,Nathan, Ph.D; Bryce, Graeme, Ph.D; Cantor,Elinor, Ph.D; Cardinale,George, Ph.D; Cohen, Bennett Ph.D; Coleclough, Christopher Ph.D; Conney, Allan Ph.D; Connor,John A. Ph.D; Curran, Tom Ph.D; Dairman, Wallace Ph.D; Dalton, Stephen Ph.D; Denckla, W. Donner Ph.D; DePamphilis, Melvin L. Ph.D; Drzeniek, Rudolph Ph.D; Ennis, Herbert L. Ph.D; Furuchi,Yasuhiro Ph.D; Gage, L. Patrick Ph.D; Gautier, Jean Ph.D; Gibson, Kenneth D.Ph.D; Gordon, Adrienne Ph.D; Greenspan, Ralph J. Ph.D; Gridley, Thomas H. Ph.D; Hedgecock, Edward M.Ph.D; Horecker, Bernard L. Ph.D; Horn, Richard Ph.D; Howard, Kenneth R. Ph.D; Hutton, John J. Jr. M.D; Kaback, H. Ronald M.D; Kerwar, Suresh Ph.D; Koch, Gebhard J. M.D; Kung,Hsiang-fu Ph.D; Levin, Wayne M.S; Lieberman, Paul Ph.D; Lu, Bai Ph.D; Maeda, Shuichiro M.D; Margolis, Frank Ph.D; Margolskee, Robert F. M.D, Ph.D; McAuslan,

Brian R. Ph.D; McMahon, Andrew P.Ph.D; Miller, David Ph.D; Montal, S. Mauricio M.D.,Ph.D; Monahan, John Ph.D; Morgan, James Ph.D; Nelson, Nathan Ph.D; Neurath, Claire Ph.D; Newman, Michael J. Ph.D; Nuss, Donald L. Ph.D; Ochoa, Severo, M.D, Ofengand, Edward James Ph.D; Pestka, Sidney M.D; Reddy, Premkumar Ph.D; Reeves, John Ph.D; Rosen, Craig Ph.D; Rushlow, Christine Ph.D; Sachs, Howard Ph.D; Sato Narukoku T. Ph.D; Marie Ph.D; Shatkin, Aaron J. Ph.D; Sherman, Michael I. Ph.D; Siddiqui, Mohammad Ph.D; Skalka, Ann, Ph.D;Spector, Sydney Ph.D; Stacey, Dennis Ph.D; Stein, Stanley Ph.D; Stewart, Colin L. Ph.D; Thomas, Paul Ph.D; Tsolas, Orestes Ph.D; Udenfriend, Sidney Ph.D; Wassarman, Paul Ph.D; Webb, David R. Ph.D; Weissbach, Arthur Ph.D; "Weissbach, Herbert Ph.D; Wood, Alexander, Ph.D; Woychik, Nancy Ph.D;

Adjunct Members

Burns, John J. Ph.D; Kuntzman Ronald Ph.D; Riker, Walter F. Jr. M.D; Spiegelman, Sol Ph.D.

Administrative Director

Snyder,Richard L. B.A.

Administrative Staff

Bartl, Paul, Ph.D; Cardinale, George J. Ph.D Eilert, William, Ph.D; Grantham, John; Galioto, Olga; Nuss, Christa L. Ford, Sherry A.

Research Scientists

Abbondanzo, Susan M.Sc; Bekesi, Eva; Bhatt, Harshida M.S; Bolden, Arthur M.S; Brink, Larry B.S; Caldwell, Paul T. M.S; Chang, David M.S; Chang, Richard M.S; Chen, Chei n-Hwa B.S; Fimiani, Carolyn B.A; Gerber, Louise M.S; Grillo, Mary M.S; Hellmann, Wilhemine B.S; Hempstead, James B.A; Huang, Mon-Tuan Ph.D; Katz, Adriana; LaFiandra, Alba B.S; Luk, Daniel M.S; Maguire, Maureen T. M.A; Mandiyan, Sreekala V. M.S; Mauser, Ljubica M.S; Maxon, Mary; McKune, Keith J. B.S; McMahon, Jill A. M.S; Miranda, Miriam B.S; Morgan, Maureen B.A; Martillo, Steven B.S; Moschera,John Ph.D; Nelson, Hannah M.Sc; Nurse, Kelvin B.S; Patel, Lekha M.S; Pawlyk, Diane M.S; Redfield, Betty B.S; Ryan, Dene M.S; Semenuk, Gloria M.S; Shaw, David R. M.S; Siekierka, John Ph.D; Skaleris, Diane M.S; Spears, Carlos M.S; Sun, Shaiu C. B.S; Verselis, Lynne, M.A; Ward, Cheryl M.S; Wurzburger, Robert M.S;Yamasaki, Edith.

Distinguished Visiting Scientists

Avron, Mordhay, Ph.D; Galun, Esra, Ph.D; Hinkle, Peter C. Ph.D; Littauer, Uriel Z. Ph.D.

Visiting Scientists

Abbott, Mitchel T. Ph.D; Altman, Arie Ph.D; Artman, Michael Ph.D; Bellamy, Alfred R. Ph.D; Benaventa, Javier Ph.D; Bhaduri, Amarnath Ph.D; Blatt, Cila Ph.D; Cancedda, Ranieri M.D; Chang, Wen-Chang Ph.D; Cohn, Dunell E. Ph.D; Datta, Asis Ph.D; Datta, Asoke Ph.D; Datta, Kasturi Deshpande Amrut Ph.D; Dhillon, Tarlochan S. Ph.D; Dickerman, Herbert W. M.D; Dunestone, John Ph.D; Eisenberg, Max Ph.D; Ejiri, Shin-irchiro Ph.D; Fantozzi,Roberto M.D; Filipowicz, Witold M.D; Fox,J. Eugene Ph.D; France-Fernandez, Maria Theresa Ph.D; Friedberg, llan Ph.D; Froesch, E. Rudolph M.D; Fulbright, Dennis W. Ph.D; Fuller, Goerge C. Ph.D; Furuichi, Vasuhiro Ph.D; Gambale, Franco; Garcia, M. Luisa Ph.D; Geffen, Laurie B. Ph.D; Gentz, Reiner L. Gershon David Ph.D; Getchell, Thomas V. Ph.D; Gilvarg, Charles Ph.D; Gornicki, Piotr; Grau, Fredrick H. Ph.D; Green, Michael Ph.D; Greenblatt, Charles M.D; Gutnick, David, Ph.D; Haehnel, Wolfgang H. Ph.D; Halpern, Yeheskel Ph.D; Hay, John Ph.D; Heller, Michael Ph.D; Hollyday, Margaret Ph.D; Horinishi, Hirao Ph.D; Hatta, Kunimoto Ph.D; Imai, Mitsunobu Ph.D; Iwata, Kazushi Ph.D; Iwata, Kazushi Ph.D; Jarratt,

Bevyn Ph.D; Kaplan, Nathan Ph.D; Katz, Edward Ph.D; Kessler, Bezalel Ph.D; Khan, Fazlur Ph.D; Kindler, Samuel Ph.D; Kolarav, Jordan Ph.D; Kolesnikov, Stanislav Ph.D; Komiyama, Tadazumi Ph.D; Kopecky, Jan Ph.D; Kotler, Moshe Ph.D; Krauskopf, Manuel Ph.D; Krzyziosiak, Wlodzimierz Ph.D; Legocki,Andrzej Ph.D; Liao, Margaret Ph.D; Lichstein, David Ph.D; Lui, Xin-Yuan; Meza-Basso, Luis Ph.D; Millard, Sara Ph.D; Millard, Sara Ph.D; Mishra, Ram K. Ph.D; Nagatsu, Toshiharu M.D; Nakamoto, Hitoshi Ph.D; Nasgatsu, Toshiharu M.D; Ngai, S.H M.D; Padan, Etana Ph.D; Padan, Etena Ph.D; Pai, Anna Ph.D; Pai, Anna Ph.D; Panneerselvam, Chinnakkannu Ph.D; Radhakrishnan, Amuthur N. Ph.D; Radsak, Klaus Ph.D; Rajamanickam, Chellam; Riker, Walter M.D; Rossier, Jean-Pierre M.D., Ph.D; Rottenberg, Hagai Ph.D; Rubinstein, Mencachem Ph.D; Savarese, John J. Ph.D; Schuldiner, Shimon Ph.D; Sela, llan Ph.D; Singh, Vishwa Nath Ph.D; Smith, David J. D.D.S; Sorensen, Craig Ph.D; Steinmetz, Phillip R. M.D; Sulser, Fridolen M.D; Szabo, Gabor M.D; Szvoboda, Gyorgy Ph.D; Taiz, Lincoln Ph.D; Tan,Shencaco; Tate, Suresh S. Ph.D; Then, Rudolf Ph.D; Tolbert, Bert Ph.D; Tsolas, Orestes; Twardowski, Tomasz Ph.D; Urade,Yooshihiro; Vince, Robert Ph.D; Wahlstorm, Agneta Ph.D; Walz, Dieter; Weather, Jean M. Ph.D; Williams, John F. Ph.D; Xu, Gen-Jun Mr; Zhu, Xin-liang M.D.

Research Fellows/Guest Workers

Akao, Teruaki,Ph.D; Almgren,Olle, M.D; Armentrout, Richard, Ph.D; Arnold, Hans-Henning,Ph.D; Aubertin, Anne-Marie, D.S; Ph.D; Ben-Porath, Edna, Ph.D; Berger, Joel,Ph.D; Bilder, Glenda, Ph.D; Bleil,Jeffrey, Ph.D; Bolder, Glenda, Ph.D; Bolla, Robert, Ph.D; Brown, Charles, Ph.D; Brun, Gilbert, Ph.D; Ph.D; Buff, Klaus, Ph.D; Cancedda, Fiorella, Ph.D; Cancedda, Ranieri, Ph.D; Cantor, Jerome, M.D; Carrasco, Nancy, M.D; Cavalieri, Ralph, M.D; Chatterjee, Nando, Ph.D; Chen, Gregory, Ph.D; Chinali, Gianni, Ph.D; Cohn, Dunell, Ph.D; Colonna, Richard, Ph.D; Crane, Laura, Ph.D; Crivellaro, Omar, Ph.D; Crivellaro, Omar, Ph.D; Deshpande, Amrut, Ph.D; Dzugaj, Andrzej, M.D; El-Dorry, Hamza, F. Ph.D; Eskin, Barnet, Ph.D; Fasy, Thomas, M.D; Fasym, Thomas, M.D; Filipowics, Witold, M.D; ;Finberg,John, Ph.D; Garel,Jean-Pierre, Ph.D; Gerben, Craig R.Goidl,Jo Alene, Ph.D; Graves, Peter, Ph.D; Hachmann, John P.Ph.D; Han, Zheng-Tao, M.S; Haung, Kuri-huang, Ph.D; Hatta, Kunimoto Ph.D; Hayakawa, Taro, Ph.D; Hermoso, Jose, Ph.D; Hong,Jen-shiang, Ph.D;; Horinishi,Hiroo, Ph.D; Horowitz, Peter, Ph.D; lwatsuki,Kasuhiko, Ph.D; Jetten, Anton, Ph.D; Jilka, Robert, Ph.D; Kamar, S. Anand, Ph.D; Kawashima, Koichiro, Ph.D; Kerwar, Koeppe,John,Ph.D; Kozak, Marilyn, Ph.D; Kozak, Marilyn, Ph.D Kumar, Ashok, Ph.D; Langness, Uwe, M.D; Lavers, Gene, Ph.D; Lavi,Sara, Ph.D; Lee, Young Joo, M.D; Lubzens, Esther, Ph.D; Manen, Carol-Ann, Ph.D; Martin, Jose L. Ph.D; McGee, James O'D. M.D.; Mendez, Enrique, Ph.D; Miller, Ronald, Ph.D; Mittelstadt, Rudolph, Ph.D; Moon, Hong Mo, Ph.D; Moyer, Sue D.,Ph.D; Muthukrishnan, Subbaratnam, Ph.D; Nagatsu, lkuko, Ph.D; Nakai,Noburu, Ph.D; Nishikimi, Morimitsu, M.D; Nissen, Reinhart, M.D; Nombella, Cesar, Ph.D; Nureddin, Aida, Ph.D; O'Brien, Robert, Ph.D; Oberg,Bo, Ph.D; Ooshima, Akira, M.D;

Osheroff, Phyllis, Ph.D Padrali-Noy, Guido, Ph.D; Pantuck, J. Eugene, M.D; Paoletti,Francesco, Ph.D; Parker, Robert J. Ph.D; Pearson, David, Ph.D ;Pliska, Vladimir, Ph.D; Poonian, Mohindar S. Ph.D; Porter, J. Steven, Ph.D; Ramagopal,Sabbanaidu, Ph.D; Ray, Usharanjan, Ph.D; Rosenfeild, Henry, Ph.D; Sasaki,Takashi,Ph.D; Schuldiner, Shimon, Ph.D; Shin, Sean H.,Ph.D; Shoeman, Robert, Ph.D; Short, Steven, Ph.D; Sierra, Jose, Ph.D; Soga, Jose M., Ph.D; Spadari,Silvio, Ph.D; Spiess, Eberhard, Ph.D; Stroobant, Paul,Ph.D; Trajkov, Trajko, M.D; Van Den Berg,J.W.O., Ph.D; Yaffe, Lyn, M.D;Yamada, Michiyuki,Ph.D; Yamamoto, Itaru,Ph.D; Zahavi, Tova, Ph.D.

Postdoctoral Fellows

Abarzua, Patrica Ph.D; Abate, Corrine T. Ph.D; Abo, Masonabu Ph.D; Abraham, Gordon Ph.D; Abraham, Shiomo Ph.D; Abrams, Beatrice Ph.D; Aceto, Joseph F. Ph.D; Afar, Ronith Ph.D; Aguanno, Ann M. Ph.D; Ahmed, Aftab Ph.D; Akao, Teruaki Ph.D; Akinyosoye, Olufemi Ph.D; Alder, Richard Ph.D; Alexander, Fern Ph.D; Alexander, Witiko R. M.D; Alicivar-Warren, Acacia Ph.D; Amir-Shapria Dekel Ph.D; Amthauer, Rodolfo M. Dr; Anderson, David M. Ph.D; Arnold, Hans-Henning Ph.D; Asano, Tomiko Ph.D; Audhya, Tapan Ph.D; Avenia, Richard Ph.D; Bailey, Carole A. Ph.D; Baker, Suzanne J. Ph.D; Bakin, Andrey Ph.D; Balestreri, Ettore Ph.D; Bandiera, Steivio Ph.D; Bandiera, Stelvio Ph.D; Barbone, Francis P. Ph.D; Barcia, Miguel Ph.D; Barkai, Uriel Ph.D; Barkey, Ronnie Ph.D; Barnes, Eugene Ph.D; Baumeister, Hans Ph.D; Baxt, Barry Ph.D;

Behn, Christian A. Ph.D; Belan, Pavel V. Ph.D; Beltran, Carmen Ph.D; Berger, Joel Ph.D; Berkowitz, Barry Ph.D; Bettini, Ezio Ph.D; Betzer, Andreas G. Ph.D; Biffo, Stefano Ph.D; Bilder, Glenda Ph.D; Bloom, Mark Ph.D; Bloomstein, Michael Ph.D; Blue, William Ph.D; Blume, John E. Ph.D; Bolder, Glenda Ph.D; Boone, Lawrence Ph.D; Botelho, Lynne Ph.D; Both, Gerald Ph.D; Both, Gerald, Ph.D; Brandsch, Roderich Ph.D; Breindl, Michael Ph.D; Brun, Gilbert Ph.D; Brunet, Lisa J. Ph.D; Buccafusco, Jerry Ph.D; Buiakova, Olesia Ph.D; Burhans, William C. Ph.D; Burrus, Laura Ph.D; Buttyan, Ralph Ph.D; Campbell-Ratliff, Jeanne Ph.D; Cancedda, Fiorella Ph.D; Canfeild, Victor A. Ph.D; Cannon, Gordon Ph.D; Cantor, Eilnor Ph.D; Carey, Alisoun H. Ph.D; Carlino, Anthony Ph.D; Carrasco, Nancy M.D; Carter, Carol Ph.D; Catz, Diana S. Ph.D; Centiempo, Yves Ph.D; Ceruzzi, Marion Ph.D; Chahine, Mohamad Ph.D; Chanda, Pranab Ph.D; Chang, Chin-Hai Ph.D; Chang, De-Maw Ph.D; Chang, Kun-Sang Ph.D; Chang, Michael Ph.D; Chang, Nancy Ph.D; Chattopadhyay, Dhubajyoti Ph.D; Chen-Kiang, Selina Ph.D; Chen, Chong-Maw Ph.D; Chen, Jane-Jane Ph.D; Chen, Jie Ph.D; Chen, Shih-Min Ph.D; Chen,Shu-Cheng Ph.D; Cheon, Joo-Wha Ph.D; Chernaya, Galina Ph.D; Chitnis, Parag R. Ph.D; Choi, Gill H. Ph.D; Chow, Nang-Ly Ph.D; Chu Fredrick K. Ph.D; Chu, Douglas K. Ph.D; Ciesiolka, Jerzy Ph.D; Citerella, Ronald Ph.D; Clarke, Catherine Ph.D; Cochrane, Alan W. Ph.D; Cohen, Donna R. Ph.D; Cohen, Marlene Ph.D; Cohn, Bennett Ph.D; Coleman, Timothy A. Ph.D; Colonna, Richard Ph.D;

Colonna, Richard Ph.D; Condresu, C. Madalina Ph.D; Cansler, Thomas G. Ph.D; Coppola, George Ph.D; Counts, David, F. Ph.D; Cox, G. Stanley Ph.D; Crane, Laura Ph.D; Craven, Mark G. Ph.D; Cribbs, Leanne Ph.D; Crivellaro, Omar Ph.D; Cullinan, Emily B. Ph.D; Cunningham, Phlip R. Ph.D; D'Arcanglo, Gabriella Ph.D; Dahmen, Norbert J. M.D; Dalie, Barbara L. Ph.D; Dall, David J. Ph.D; Danciger, Efrat Ph.D; Danciger, Efrat Ph.D; Danishefsky, Kenneth Ph.D; Darzynkiewicz, Edward Ph.D; Dasgupta, Purander Ph.D; Datta, Kasturi Ph.D; David, Jean-Claude Ph.D; De, Bishnu Ph.D; Delaney, Pamela Ph.D; Denman, Robert B. Ph.D; Deutsch, Victor Ph.D; Devine, Evelyn; Dickinson, Michael H.m Ph.D; Dillon, Patrick J. Ph.D; Doller, Elizabeth Ph.D; Doller, Herbert J. Ph.D; Dominguez, Fernando Ph.D; Donnerer, Josef M.D; Dottavio, Diane Ph.D; Drabkin, Harold Ph.D; Duffy, Lawrence Ph.D; Durkin, John T. Ph.D; Durnford, Joyce Ph.D; Eggen, Kathleen Ph.D; Eglitis, Martin Ph.D; Ehelard, Yann Ph.D; Enquist, Lynn W. Ph.D; Erikson-Vitanen, Susan Ph.D; Erion, Jack Ph.D; Ernst, Heinrich Ph.D; Evinger, Marion Ph.D; Ezra, Elhanan Ph.D; Ezra, Elhanan Ph.D; Fairfeild, Stephen Ph.D; Fallon, Anthony Ph.D; Faria, Juarez Braz de Ph.D; Fasy, Thomas M.D; Fernstrom, John D. Ph.D; Ferreira, Paulo C. Ph.D; Ferreria, Paulo C. Ph.D; Fervuer, Jean-Francois Ph.D; Filipowicz, Aleksandra Ph.D; Filipowicz, Witold M.D; Finberg, John Ph.D; Fiorica, Elena Ph.D; Fisher, Linda E. Ph.D; Fleming, Kenneth Ph.D; Fleminger, Gideon Ph.D; Fliss, Henry Ph.D; Floor, Erik Ph.D; Forrest, Douglas

Ph.D; Foster, David Ph.D; Fox, Lawernce Ph.D; Franco del Amo, Francisco Ph.D; Freire-Moar, Jose M. Ph.D; Freire, Manuel Rama Ph.D; Freund, Erwin Ph.D; Fridlender, Bertold Ph.D; Fry, Michael Ph.D; Fu, Chee-Fook Ph.D; Fujiwara, Yuko Ph.D; Gacheru, Steven N. Ph.D; Gadot, Michal Ph.D; Ganguly, Kalyan Ph.D; Garattini, Enrico Ph.D; Garber, Richard Ph.D; Garfinkle, Barry Ph.D; Garica, M. Louisa Ph.D; Garret, Maurice Ph.D; Gavin, Brain J. Ph.D; Ge, Ruowen Ph.D; Gerassi, Esther Ph.D; Giantini, Michael Ph.D; Gilbert, David M. Ph.D; Giles, Penelope Ph.D; Gill, Dalip Singh Ph.D; Gilligan, James Ph.D; Gintzler, Alan Ph.D; Giorda, Roberto Ph.D; Giri, Judith Ph.D; Giri, Judith Ph.D; Gitman, Adolfo Ph.D; Gold, Paul Ph.D; Goldberg, Gregory Ph.D; Goldkorn, Tzipora Ph.D; Gonzalez-Martinez, Marco T. Ph.D; Goodall,Gregory Ph.D; Goodall, Gregory J. Ph.D; Goodall, Yee-Sim K. Ph.D; Goorha, Rakesh Ph.D; Goren, Tamar Ph.D; Gornicki, Piotr Ph.D; Gosule, Leonard Ph.D; Graves, Peter Ph.D; Green, Michael Ph.D; Greene, Richard A. Ph.D; Greenstein, Michael Ph.D; Greve, Jeffery Ph.D; Griffith, Leslie Ph.D; Grimberg, Jacob Ph.D; Grosfeld, Haim Ph.D; Gross, Stephen R. Ph.D; Gruber, Kenneth A. Ph.D; Gulber, Ueli Ph.D; Gutierrez, Anthony G. Ph.D; Gutierrez, Crisanto Ph.D; Haberman, Frank Ph.D; Hale, Calvin Ph.D; Hall, Alan Keith Ph.D; Hannappel, Ewald Ph.D; Harding, Joseph W. Ph.D; Harikumar, Padmanabha Ph.D; Haritos, Athanassios Ph.D; Haro, Cesar de Ph.D; Haran, Jay Ph.D; Harper, Mary Ph.D; Harris, Raymond Ph.D; Head, Richard J. Ph.D; Healy,

Cynthia Ph.D; Heine, Uwe Ph.D; Heinhorst, Sabine Ph.D; Hermosa, Jose M.Ph.D; Hernandez, Arturo Ph.D; Herz, Ruth Ph.D; Herzlinger, Doris Ph.D; Hidaka, Hiroyoshi M.D.,Ph.D; Hillbush, Brian S. Ph.D; Hirsch, James D. Ph.D; Hirsch, Michael Ph.D; Hiscott, John Ph.D; Hishizawa, Tokutaro M.D.,Ph.D; Hjelle, J. Thomas Ph.D; Ho, Ching Ph.D; Hoal, Eileen Ph.D; Hobbs, Donna Ph.D; Hong, Su-Chen Ph.D; Hopwood, Blair Ph.D; Horak, Pavel M.D, Ph.D; Horinishi, Hiroo Ph.D; Howard, Andrew D. Ph.D; Howells, Richard Ph.D; Howland, Richard Ph.D; Hsu, Lillian Ph.D; Hua, Jia-Cheng Ph.D; Huang, Kuei-huang Ph.D; Hull, Melissa W. Ph.D; Hullihan, John Ph.D; Hynes, Martin Ph.D; Imai, Kazuhiro Ph.D; Inamine, Gordon Ph.D; Infante, Diogenes J. Ph.D; Innis, Michael A. Ph.D; Isaacs, Charles Ph.D; Ishii, Naoaki Ph.D; Iwatsuki,Kasuhiko Ph.D; Jaglarz, Mariusz Ph.D; Jamieson, Andrew T. Ph.D; Jerez, Carlos Ph.D; Jetten, Anton Ph.D; Jilka, Robert Ph.D; Jilka, Robert Ph.D; Johnson, Mark Ph.D; Jones-Held, Susan L. Ph.D; Jones, Barry Ph.D; Jones, Kathryn S. Ph.D; Ju, Grace Ph.D; Junghans, Richard P. Ph.D; Jursky, Frantisek Ph.D; Kaczorowski, Gregory, Ph.D; Kamely, Daphne,Ph.D; Kaneko, Kataro,Ph.D; Kang,;Hyen-Sam,Ph.D; Kanke-Zur, Mordechai,Ph.D; Kantrowitz, Joel, Ph.D; Kao,Teh-hui, Ph.D Karjainen, Klaus, Ph.D, M.D Kaspar, Petr, Ph.D" Katarova, Zaya, Ph.D; Katz, Richard A. Ph.D; D Kavety, Balaji K.,Ph.D; Kawano, Teruaki, M.D; Kawashima, Koichiro, Ph.D; Kelley, Michael, V.M.D; Kelly, Li, Ph.D; Kelly, Rosemaria, Ph.D; Kenessey, Anges, Ph.D; Kerppola, Tom K., Ph.D;

Kessler, Dana J. Ph.D; Khan, Fazlur Rehman, Ph.D; Khandekar, Pramod, Ph.D; Kiement, John K., Ph.D; Killian, Patrica, Ph.D; Kilpatrick, Daniel, Ph.D; Kim, Jeong, Ph.D; Kimura, Sadao, Ph.D; Kimura, Sadao, Ph.D; Kinloch, Ross, Ph.D; Kirov, Nikolai C., Ph.D; Knopf, Karl-Werner, Ph.D; Koch, Michael,Ph.D; Kodaria, Hawuyoshi, Ph.D; Kodukula, Krishna, Ph.D; Kohler, Constance, Ph.D; Kojima, Kohichi,Ph.D; Kolb, A lfred, Ph.D; Kopchick, John, Ph.D; Korbin, Barry, Ph.D; Korn, Stephen J., Ph.D; Kouri, Richard,Ph.D; "Kowalski, Steven, Ph.D; Kozak, Marilyn, Ph.D; Kraiselburd, Edmundo, Ph.D; Krauskopf, Manual R. Ph.D; Kream, Richard, Ph.D; Kroath, Hans ,Ph.D; Kudrycki, Katarzyna, Ph.D; Kunsch, Charles, Ph.D; Kurschner, Corelia, Ph.D; Kuttan, Ramadasan, Ph.D; La Colla, Paola, Ph.D; Labdon, James E., Ph.D; Lai, Fong M. Ph.D; Lambert, Judith, Ph.D; Langer-Safer, Pennina, Ph.D; Langer-Safer, Pennina, Ph.D; Langer, Jerome, Ph.D; Larson, Thomas G., Ph.D; Lau, Michele, Ph.D; Lavi, Sara, Ph.D; Lavicky, Jan, Ph.D; Lavu,Sukadev, Ph.D; "Lazo, Pedro Sanchez, Ph.D; LeBlanc, Girard, Ph.D;Lee, Arianna, Ph.D; Lee, Chung S., Ph.D; Lee, Jonathan, Ph.D; Lee, Yonih, Ph.D; Lee, Young . Ph.D; Lehm, Hans-Werner, Ph.D; Lessard,James L. Ph.D; Leung, Wing-Nang, Ph.D; Leustek, Thomas, Ph.D; Levinson, Susan L. Ph.D; Levy, Aharon, Ph.D; Levy, Warren, Ph.D; Lewis, Randolph, Ph.D; Li, Hui, Ph.D; Lichtshtein, David, Ph.D; ;Lievano, F. Arturo, Ph.D; Lill, Holger, Ph.D; Lin, Fwu-Lai, Ph.D; Lin, Gene W. Ph.D; Lin, Wan-Lu, Ph.D; Linden, David J. Ph.D; Lira, Sergio

A. Ph.D; Litscher, Eveline, Ph.D; Liu, Chen K., Ph.D; Liu, Chengyu, Ph.D; Liu, Qing-Rong, Ph.D; Lolkema, Julius S. Ph.D; Lombardi, Frank J. Ph.D; Long, Zhifeng, Ph.D; Longiaru, Mathew, Ph.D; Lopez-Corcuera, Beatriz Ludwig, Gerald, Ph.D; Lugo, Delia I. Ph.D; Mace, David, Ph.D; MacGregor-Leon, Pascale F. Ph.D; Machida, Atsuhiko,; Ph.D; Marconi, Richard T. Ph.D; Maddux, Regina, Ph.D; Maeda, Shu-ichiri,M.D; Mahanty, Nishith K. Ph.D; Majumder, Sadham, Ph.D; Mallo, Moises, Ph.D; Mandel, Myrna, Ph.D; Mandiyan, Valsan, Ph.D; Manen, Carol-Ann, PH.D; Mann, Jeffrey, Ph.D; Manne, Veeraswamy, Ph.D; Marangos, Paul, Ph.D; Mariano, Thomas, Ph.D; Martin-Zanca,Dionislo, Ph.D; Martin, Mitchel L. Ph.D; Martinez-Salas, Encarnacion, Ph.D; Martinez-Salas, Encarnacion, Ph.D; Marzachi, Cristina, Ph.D; Marzchi, Cristina, Ph.D; Maser, Joseph, Ph.D; Masters, Paul S. Ph.D; Matsushita, Kazunobu, Ph.D; Matthaei, Klaus, Ph.D; Max, Marianna, Ph.D; Maxcy, Philip, Ph.D; Maxwell, Stephen E. Ph.D; Mccandliss, Russell, Ph.D; McClements, William, Ph.D; McCue, Peter, M.D; McDermott, Martin, Ph.D; Mcinnes, James L. Ph.D; McKelvy, Jeffrey, Ph.D; McKinnon, Peter J., Ph.D; McLaughlin, Susan K.Ph.D; Mehta, Bela A. Ph.D; Menick, Donald R. Ph.D; Mensi, Nahla, Ph.D; Miara, Joelle, Ph.D; Micanovic, Radmila, Ph.D; Micanovic, Radmila, Ph.D Miller, Ronald L.. Ph.D Misumi, Masarou, Ph.D Mitchell, Robert, Ph.D; Mitsui, Ken'ichiro, Ph.D; Mittelstadt, Rudolf, Ph.D; Miyazawa, Hiroshi,Ph.D; Moiyama, Yoshinori, Ph.D; Molinar-Rode, Ricardo A. Ph.D; Moller, C. Christopher, Ph.D;

Moore, Paul A., Ph.D; Moriyama, Yoshinori, Ph.D; Morriessey, Jeremiah, Ph.D; Morris, Claire E. Ph.D; Moschera, John, Ph.D; Mosckovitz, Rachel, Ph.D; Moskovitz, Jackob, Ph.D; Mulcahy, Linda, Ph.D; Muroff, Ivor, Ph.D; Nadi, Suzan, Ph.D; Nakai, Noboru, Ph.D; Nakashima, Kunio, Ph.D; Nalin, Carlo M. Ph.D; "Nalin, Carlo, Ph.D, Nallaseth, Ferez S., Ph.D; Narula, Satwant, Ph.D; Nash, Barbara, Ph.D; Nash, Barbara, Ph.D; Natale, Darren, Ph.D; Natalini, Paola, Ph.D; Natalini,Paolo, Ph.D; Negre, Didier M. Ph.D; Nellboeck-Hochstetter, Peter A. Ph.D; Ng, Lily, Ph.D; Nicholson, Diarmuid, Ph.D; Niederman, Robert A.,D.V.M., Ph.D; Nishikawa, Takashi, Ph.D; "Noe, Mariadele, Ph.D; Nolan, Joseph, Ph.D; Norgard, Michael, Ph.D; Nothias, Jean-Yves, Ph.D; Noumi,Takata, Ph.D; Oberdick, John D. Ph.D; Ohana, Bella, Ph.D; Ohmachi,Tetsuo, Ph.D; Oiki, Shigetoshi, M.D; Oka, Kazuhiro, Ph.D; Okuyama, Akira, Ph.D; Olafesson, Petur, Ph.D; Olsen, Henrik S. Ph.D; Ooshima, Akira, M.D; Oostergetel, Geritt, Ph.D; Opperman, Hermann, M.D; Oppi, Cristina, Ph.D; Orgad, Sara, Ph.D; Ortega, Jose M. Ph.D; Osses, Luis R. Ph.D; Oste, Christian, Ph.D; Otani, Satoru, Ph.D; Ozer, Josef, Ph.D; Palese, Peter, Ph.D; Pan, Lu-xing, Ph.D; Paoletti, Francesco, Ph.D; Parekh, Trilok V. Ph.D; Parr, Brian A. Ph.D; Paul, Palash R. Ph.D; Pearson, David, Ph.D; Pennica, Diane, Ph.D; Periasamy, Muthu G. Ph.D; Perler, Francine, Ph.D; Peskar, Bernhard, Ph.D; Peterson, Robert, Ph.D; Petrozzino, Jeffery J. Ph.D; Pong, Shen-Shung, Ph.D; Popieniek, Paul H., Ph.D Porta, Angela, Ph.D" Potor, J. Steven, Ph.D; Pozzo-Miller,

Lucas D. Ph.D; Preuss, Axek K. Ph.D; Prezioso, Girolamo, Ph.D; Pugatsch, Thea, Salorta, Poonam, Ph.D; Purvis, David J., Ph.D; Puttner, Irene, Ph.D; Racevskis, Janis, Ph.D; Rahman, Mohammad A. Ph.D; Rama, (Freire), Manuel, Ph.D; Ramagopal, Subbanaidu, Ph.D; Ramalingam, Ramachandran, Ph.D; Ramalingam, Sandhya, Ph.D; Ramon, Eliezer, Ph.D; Ramos, Sophia, Ph.D; Rashidbaigi,Abba, Ph.D; Rauscher, Frank J., Ph.D; Reenstra, William, Ph.D; Reilly, Patrica, Ph.D; Rein, Rita S.Ph.D; Reisfeld, Avi, Ph.D; Reuben,Roberta, Ph.D; Richardson, Michael, Ph.D; Richter, Hans, Ph.D; Rimon, Gilad, Ph.D; Rioux, Francis, Ph.D; Robakis, Nikolaos, Ph.D; Robertson, Dan, Ph.D; Robichon, Alain, Ph.D; Rodriguez, Monica B. Ph.D; Roepe, Paul D. Ph.D; Rogers-Slemmon, Katherine, Ph.D; Rogers, Christina M. Ph.D; Rosiere, Thomas K. Ph.D; Rosson, Dan, Ph.D; Roth, Hans E. Ph.D; Roy, Paul H. Ph.D; Roy, Paul, Ph.D; Ruben, Steven M. Ph.D; Rubinstein, Menachem, Ph.D; Rudnick, Gary, Ph.D; Ruggieri,Silverio, Ph.D; Ruiz-Seller, Betina M. Ph.D; Rush, Robert A. Ph.D; Saborio, Jose L. M.D; Saborio, Jose, M.D; Sadhale, Parag, Ph.D; Saidapet, Chandrika, Ph.D; Sakai, Masao, Ph.D; Sakai,Yutaka, Ph.D; Salomon, David S. Ph.D; Sambucetti, Lidia C. Ph.D; Samejima, Keijiro, Ph.D; Samols, David, Ph.D; Samuels, Fred, Ph.D; Sanchez, M. Angeles, Ph.D Sanchez, M. Augeles, Ph.D; Sangamesawaren, Lakshimi, Ph.D; Sapiro, Jacob, Ph.D; Sapiro, Katherine, Ph.D; Sarker, Hemanta K.Ph.D; Sarker, Hemanta, Ph.D; Sarubbi, Donald J. Ph.D; Sasaki, Takashi, Ph.D; Sato,

Paul,Ph.D; Schickler, Michael, Ph.D; Schilling, Karl, M.D; Schindler, Joel, Ph.D; Schlabach, Abner J. Ph.D; Schmid, Barbel, Ph.D; Scholl, David, Ph.D; Schrewe, Heinrich K., Ph.D; Schuld i ner, Shimon, Ph.D; Schurdak, Ma rk E. Ph.D; Schwartz, I ra, Ph.D; Seidei, Bon nie, Ph.D; Sellens, Ma rti n, Ph.D; Shai n berg, Asher, Ph.D; Shapria, Roni, Ph.D; Shea, Ma rti n J. Ph.D; Shen, Liya, P h.D; Shorr, Jeosh ua, Ph.D; Short, Steven, Ph,D; Sied lecki, Ja n usz, Ph.D; Sied lecki, Jan usz, Ph.D; Sierra, Jose M., Ph.D; Sigiu ra, Motota ka, M.D; Silverma n, Rachel C. Ph.D; Silverma n, Robert, Ph.D; Singh, Bal raj, Ph.D; Skelly, Susan M. Ph.D; Slammon, J ohn R. Ph.D; Slaughter, Robert S. Ph,D; Slemmon, John R. Ph.D; Sloma, Ala n, Ph.D; Smart, David, Ph.D Smart, Robert C. Ph.D; Smeyne, Richard J. Ph.D; Smith, David E. Ph.D; Smith, David, Ph.D; Smith, Mark, Ph.D; Smith, Robert, Ph.D; Soares, Holly, Ph.D; Sahel, lndria, Ph.D; Solitis, Daniel A. Ph.D; Soneberg, Nahum, Ph.D; Sonnenberg, June L. Ph.D; Sonnenfeld, Kenneth, Ph.D; Stark, Kevin L. Ph.D; Stassen, Francis L. Ph.D; Stefanos, Simon, Ph.D; Stein-Izsak, Cathy, Ph.D; Stein, Jill C. Ph.D; Stein, Stanley, Ph.D; Stein ke, Laurey A. Ph.D; Stern, Alvin, Ph.D; Stern, Alvi n, Ph.D Stern, Bria n, Ph.D" Stoltzfus, C. Martin, Ph.D; Stram, Yeduda, Ph.D; Stram, Yehuda, Ph.D; Stroobant, Paul, Ph.D; Suda, Hiroyuki, Ph.D; Sugrue, Stanley, Ph.D; Sulner, Joseph, Ph.D; Supek, Frantisek, Ph.D; Supekova, Lubica, Ph.D; Swiatek, Pamela J. Ph.D; Swinney, David C. Ph.D; Sydor, Wasyl, Ph.D; Sydor, Wasyl, Ph.D; Szabo Gabor, M.D; Szvoboda, Gyorgy, Ph.D; Tahara, Stanely,

Ph.D; Tahara, Stanley, Ph.D; Taisky, Michael, Ph.D; Takada, Shinji, Ph.D; Takasaki, Yozo, Ph.D; Takashima, Hideo, Ph.D; "Taketo, Makoto, M.D., Ph.D; Talib, Schei, Ph.D; Tamburrni, Maurizo, Ph.D; Tamura, Shigehiko, Ph.D; Tan, Kong Beng, Ph.D; Taniguchi, Takashi, Ph.D; Tarnowski, S. Joseph, Ph.D; Tartaglia, James, Ph.D; Teirelbaum, Zvi, Ph.D; Tejwani, Gopi A. Ph.D; Testa, Douglas, Ph.D; Thomas, Edwin, Ph.D; Thomas, Paul, Ph.D; Thronton, George,Ph.D; Tocci, Michael, Ph.D; Todhunter, John A. Ph.D; Tokuda, Hajime, Ph.D; Toledo, Hector M. Ph.D; Treadwell, Benjamin, Ph.D; Trumble, William, Ph.D; Trumble, William, Ph.D; Tsai, Men-Hwei, Ph.D; Tscherne, Joan, Ph.D; Tscherne, Joan, Ph.D; Tucker, Philip, Ph.D; Tumminia, Santo, J. Ph.D; Turck, Christopher, Ph.D; Twardowski, Tomasz, Ph.D; Tyski,Stefan, Ph.D; Urade, Reiko, Ph.D; Urbaowski, Joseph, Ph.D; Van Heiden, Paul, Ph.D; Vania, Seppa, Ph.D; Vassilev, Lyubomir, Ph.D; Vazquez, Monica H. Ph.D; Vendrell, Monsterrat, Ph.D; Verhaagen, Joost, Ph.D; Victor, Jacob, Ph.D; Vidal, Vincent, Ph.D; Vidal, Vincent, Ph.D; Viliagines, Roland, Ph.D; Vita, Alberto, Ph.D; Vitanen, Paul, Ph.D; Wainwright, Norman, Ph.D; Walker, Anita, Ph.D; Walker, Anita, Ph.D; Walker, William, Ph.D; Walters, Eric, Ph.D; Wang, Dashou, Ph.D; Wang, Mei-tai, Ph.D; Wang, Ping, Ph.D; Wang, Shue-Yuan, Ph.D; Wang,Xin W. Ph.D; Ward, Richard, Ph.D; Warrior, Rahul,Ph.D; Weetall, Marla, Ph.D; Wei, Shu-Jing, Ph.D; Weitzman, Carl J. Ph.D; Wertheimer, Alice, Ph.D; Whitbread, Lesley, Ph.D; Wiehe, Gerhard, Ph.D; Wideman, Janusz, Ph.D Wieder,

Kenneth, Ph.D Wieder, Kenneth, Ph.D Wiekowski, Maria T., Ph.D; Wigboldus, John D. Ph.D; Wildman, David, Ph.D; Williams, Bill, Ph.D; Winger, Klaus-Jochen, M.D; Within, Steven, Ph.D; Wolfe, Richard A . Ph.D; Wong, Gwendolyn T. Ph.D; Wrzesinski,Jan, Ph.D; Wudl, Linda, Ph.D; Wydra, Robert, Ph.D; Xanthoudakis, Steven, Ph.D; Xia, Qi-Chang, Ph.D; Xie, Dian-Lin, Ph.D; Xie, Kewei, Ph.D; Xu, Zengkai, Ph.D; Yaffe, Lyn, M.D; Yamakawa, Minoru, Ph.D; Yan, Ning, Ph.D; Yang,Ning-Sun, Ph.D; Yang,Xian-Da, Ph.D; Young, Xiao-Ying, M.D; Youngleib, Gary, Ph.D; Yu, Chun Li, Ph.D; Yuzaki, Michisuke, Ph.D; Zackson, Saul L. Ph.D; ng, Maobin, Ph.D; Zhang,Qingchang, M.D; Zhao, Xiao-Xia, Ph.D; Ziai, Reza, Ph.D; Ziegler, Karl J.W. Ph.D; Zimecki, Michal, Ph.D; Zimmermann, Wolfgang, Ph.D.

Graduate Students

Anjanayaki, E.A Cai,Xiao-Yan; Chen-Kiang; Selina Dhundale; Arun Dickinson; Mary Eichner; James Fajardo; Jorge E., Griffin; Kevin J. Guo; Zongsheng Hantelle; Patrick Hayward; Michael Hill; Janet Hussain, Zamirul,M.S; Jakowiew; Sonia Kasica; Deborah, Ann Kasof, Gary Keller, Angelica Kiang,Selina; Krupen-Brown; Karen Krupen; Karen Lee, Chi-Ho Liu, Chen K. MacGregor; John MacGregor; Ana Maria, John Menezes; Elizabeth Miao; Graham G. Muller; Robert Nugnet; Matthew Oppermann; Susan Peacock; Herman Pedrosa; Fabio Ransom; John Reuben; Roberta Rhoads;

Robert E.,B.A Rhodes; Dennis Roller; Richard Salzmann; Goerge Schul; Tanya Schulz; John Siekierka; John Sipley; John Sunde; David Wing; Wag, James Weisman; David Wing;;Stuart Zarraga.

Support Staff

Alston, Enid, B.A; Andriloa; Sophe Andriola; Diane Antes; Lisa Arnold; Mary Ann Arroyo; Carmen Bobal; RoseAnn Brenner; Louise; Bridges-Snell; Arlene Byrne, Tina; Cahill; Elizabeth Cairoli; Kathy Cassidy; Carolyn Cluesman; Betsy Connelly; Jill Crane; Jonathan Cuber; Sophie Della Fave; Rita Della Fave; Rita DiPaola; Josephine Doers; Anwara Ewald; Wendy Faronea; Audrey Ford; Sherry Frasier; James Jr. Frasier; James Jr. Gailoto; Olga Galioto; Olga;"Gambacorta; Mary Ann Gatling; Barbara George; Barbara Grantham; Dion Green; Edward; Gregg, Linda J. B.A Gresham; Evelyn Hamilton; Arthur Hansen; Janet Hansen; Janet Hansford; Barbara Hernedez; Ada;Hinton; Lester; Hugenholtz; Jeron Icciardelli; Martie Jacobs; Maurice Jane; Jane Jaremkof; Kathy Jaskiewicz; Tess Johnson; Ruth Jones; Carol Jones; Kenneth Jones; Kenneth Kaminsky; Walter Kerr; Barbara Kerr; Barbara Koteles; Bela Koteles; Thersea Kozak; Irmgard Kozak; Stefan; Kruleski; Kristen B.A; Kubicz; Genowefa Kupperman; Sandra Lardieri; Ralph Leone; Rose Lewis; Peter Maffei; Lisa Marcum; Grace Marseglia; Josephine Martins; Carola McGuire; Daniel Mersereau; Diane

Michaud; Cathy, B.A; Narducci; Louise Nuss; Christa Nuss; Dana C., B.A; Oliva, Antonio Palumbo; Lori Ann Parry; Helen Pearlman; Edna Perkowski; Patrica Perry; Sharon M. Phillips; Alicia Pizzano; Carol Puyear; Mary Ann Radigan; Mari Radigan; Mari Rafter; Jo Anne Revis; Doris Rose; Tina Sauchelli; Leeann Schaffer; Theresa Schmitt; John Schroeder; David Shkolar; Olga Squirlock; Monica Tighe; Katherine Tighe; Dennis Tirone; Glauco Torres; Dot Trbovich; Anna Weintz; Robert Jr. Welborn; Carol Wright.

Animal Services

Abel; Sharon; Apruzzese; Claire Brehma; Ernest Cabral; John Cabral; John Cadogan; Henry Cupta; Petra Denny; John Donohue; Patrica Harrison; Charles Herrmann; Sharon Johnson; Kenneth Kotys; Marie Liska; Tracy Mackey; Arthur Mackey; Arthur Marrero; Edwin Nagel;Stephen Podgorsek; Frank Rizo; Jose Rizzolo; Peter Romel; Louis Sanabria; Maria Tomea; Fred White; Barbara Whyte; Malcom Wilson, James Witte, Robert.

Glassware Services

Gaiting, Margaret R.; Grantham, John E.;Supervisor Hamilton; Artie I. lnfermo; Josephine Jane; Jane Koteles; Bela Malloy; Catherine Malloy; Catherine Mandeville; Wayne M. Olivia;

Antonio Puryear; Mary A. Rafter; Jo Anne Supulveda; Welborn,Robert H.

Board of Scientific Advisors

Dr. Christian B. Anfinsen; Dr. K. Frank Austen; Dr. Julius Axelrod; Dr. J. Michael Bishop; Dr. Sydney Brenner; Dr. Donald D. Brown; Dr. Mildred Cohn; Dr. Renato, Dulbecco; Dr. Herman N. Eisen; Dr. Donald S. Fredrickson; Dr. Walter J. Gehring; Dr. Charles Gilvarg; Dr. Alfred G. Gilman; Dr. Avram Goldstein; Dr. Paul Greengard; Dr. Robert E. Handschumacher Dr. Bernard L. Horecker; Dr. Jerard Hurwitz Dr. Paul Marks; Dr. Victor A. McKusick; Dr. Bert N. LaDu Jr.; Dr. Elvin A. Kabat; Dr. Har Gobind Khorana; Dr. Norman Kirshner; Dr. Arthur Kornberg; Dr. Hilary Koprowski; Dr. Daniel E. Koshland; Dr.Joshua Lederberg; Dr. Rodolfo R. Llinas; Dr. Salvador E. Luria; Dr. Alton Meister; Dr. Daniel Nathans; Dr. Marshall Nirenberg; Dr. Mary J. Osborn; Dr. George E. Palade; Dr. Charles W. Parker; Dr. Efraim Racker; Dr. Robert G. Roeder; Dr. Saul Roseman; Dr. David Sabatini; Dr. Howard K. Schachman; Dr. J. Edwin Seegmiller; Dr. Earl R. Stadtman; Dr. DeWitt Stetten, Jr.; Dr. Charles F. Stevens; Dr. Shirley M. Tilghman; Dr. Gordon Tomkins; Dr. James D. Watson; Dr. Robert Weinberg; Dr. Norman Weiner; Dr. Charles Weissmann

CPSIA information can be obtained
at www.ICGtesting.com
Printed in the USA
LVOW04*2108171216
517762LV00001B/2/P